Linear Optimization

A Geometric Inquiry Course

Linear Optimization

A Geometric Inquiry Course

Tien Chih
Oxford College of Emory University

Tien Chih
Department of Mathematics
Oxford College of Emory University
tien.chih@emory.edu

2026 Edition
ISBN: 978-1-958469-34-7

A current version can always be found for free at
https://linopt.tienchih.org/

Cover Artist: Jean Tashima.

Student Statements

"I really liked how the textbook built up the ideas in a way that made the motivation behind each concept and theorem feel natural. The order that the material was presented in made sense, and it was clear why each new technique was introduced. I also appreciated that we often derived or proved results ourselves before seeing the formal version - it made everything fit together more logically and helped me actually understand why things work instead of just memorizing them. On top of that, the emphasis on the geometric perspective really deepened my understanding; seeing how the concepts connect visually made linear programming feel much more intuitive." - Lance Ding, Computer Science & Mathematics Student, Emory University

"I enjoyed the inquiry-based classroom experience - it transformed each class into an exciting research project. I felt a great sense of achievement when my peers and I collaborated to prove important theorems and explored how linear optimization applies to real-world scenarios. The excitement of discovery not only deepened my understanding of mathematics but also fostered lasting friendships and unforgettable memories." - Arthur Han, Mathematics & Physics Student, University of Michigan

"I would say that some of the best skills I learned from this class wasn't just the material, but the way it was taught. Doing collaborative proofs allowed for a more complete understand of not just the proof itself, but the process of coming up with proofs. The collaborative process has helped me learned how to apply knowledge from both me and my classmates to figure out principles that previously I did not know. It has helped me with research projects that I have undertaken not just in math, but in CS as well." - Jane Brennan, Computer Science & Mathematics Student, Emory University

For Tara, Byron, Max and Mya

Acknowledgements

This book is the product of multiple influences and is the by product of many threads in my life. I want to thank Steven Clontz and Drew Lewis for including me in the inaugural Team Based Inquiry Learning group. What I learned about authoring inquiry learning materials from that experience is the biggest contributor to the structure of this text. I wanted to thank the OER and PreTeXt communities, but in particular David Austin, Steven Clontz, Sean Fitzpatrick, Mitch Keller, Oscar Levin and Chrissy Safranski for their contributions not only to PreTeXt which I used to author this text, but directly help with numerous technical issues throughout the writing process. I would like to thank my brother Fong Chih for his help with the simplex pivot tool.

I would like to thank Jennifer Nordstrom and John Rock for their helpful suggestions and feedback. I would like to thank Candice Price, Miloš Savić, Jen Martindale and Jean Tashima, the whole 619 Wreath Publishing team for their support, encouragement, and genuine partnership. Their support of scholars, open access, and unique contributions is truly commendable.

Finally, I would like to thank the students, faculty and staff of Oxford College of Emory University for fostering an environment of inquiry and creativity. An especial heartfelt thank you goes to Jane Brennan, Lance Ding, Will Farthing and Ari Gurovich, the wonderful students in my first Math 346 Linear Optimization course who were taught with an embryonic version of this text, and provided valuable feedback and encouraged me to develop those notes into this text.

Our goals

This is an inquiry-learning based textbook on introductory linear optimization. Linear optimization, or linear programming, is a common but not ubiquitous course at the undergraduate level, with somewhat divergent goals and approaches. There are numerous texts for linear optimization, so it what is it that this book offers that others do not?

Linear optimization as a course serves a variety of disciplines, and can be taught in a number of different approaches. In a math curriculum, while it has linear algebra as a pre-requisite, it generally lies in the mid-high tier level of courses, prior to classes like abstract algebra and real analysis, but past calculus and linear algebra. This is a subject rich with both beautiful mathematics, as well as numerous applications to economics, computer science, operations research, sociology and other areas. So this course is an excellent choice for a variety of students, and can serve the needs of many.

The presentation of linear optimization can be computation or theory focused, and both approaches would generally be an extension of a linear algebra course which also generally has those two flavors. Either way, the material is often presented in a way that is very technical from a linear algebraic standpoint. For a student, learning the algorithms, techniques, theorems and proofs can be overwhelming. There are numerous, technically intricate arguments, and the complexity of the associated algebra can subsume the intuition behind this material, and render much of the content opaque.

I had a similar experience working with linear optimization as a graduate student, and having never taken an undergraduate linear optimization course, found myself having to play "catch-up". It didn't help that different text and resources presented the content in very different ways, and so unpacking the algorithms and theorems was a challenge. I was never satisfied with the idea that it was sufficient to memorize an algorithm or even prove the theorems, what I was looking for was an intuitive way to think about this subject, in a way that makes the results intuitive, natural, even expected.

What helped me then was recontextualizing all of the key ideas of linear optimization in geometric terms. By centering the geometry underlying each scenario, we can allow our intuition to guide things forward, and the flow of the material proceeds naturally. So this book begins with geometric realizations, translating that geometry into linear algebra, and proceeding from there to computations, but each theorem, formula and computation can be understood on this level.

Something that this approach allows is an easier entry-point for students. With geometric motivation, students have an easier time anticipating and predicting what may be true, and then this intuition can then be formalized with the appropriate statements and proofs. A friend from graduate school, Dr. Michael Severino, is an avid rock climber and gave the analogy that in mathematics, intuition was knowing where on the cliff to grab next, and rigor was the rope keeping you from falling. This book aims to present both intuition and rigor in a way that lets all students propel to the top of this cliff.

Presenting the content with this framework lends itself well to teaching this course in an inquiry based manner. Numerous studies have been done showing the efficacy of inquiry learning, and the longer term understanding students develop from learning this way. I have taught numerous courses in an inquiry based manner, and have authored inquiry learning materials, and the key to successful course is appropriate scaffolding. Presenting the right mixture of intuitive exploration, rigorous argumentation, and text or instructor intervention.

As of the writing of this text, there is a renaissance of open source technology tools aimed at visualizing, computing or demonstrating mathematical ideas in accessible, interactive ways. Use of this technology can greatly enhance the ability for students to visualize and intuit the ideas being presented in a class, or side-step tedious computations, so that students can focus their attention and time on the conceptual principles of a course.

These are the thoughts that I had in mind when I was first asked to teach linear optimization at Oxford College of Emory University. Oxford College is a small, teaching-focused, liberal arts college of Emory University of about 950 students. We teach first and second year students who, upon completing Oxford's liberal arts curriculum, proceed to Atlanta to complete their bachelor degrees. Given the inquiry driven principles of the college, and all of the above thoughts, I decided to teach this class in an inquiry manner, incorporating the ideas I've written about, and through the semester wrote a set of inquiry learning materials, which serves of the basis of the document you are currently reading.

Given the principles which informed my design choices, and the ways in which a text for an inquiry based course differs substantially from a more traditional textbook, there are some things to keep in mind as you go through this text:

- *This book is written with the assumption that students have taken a linear algebra course.* Linear algebra is the language in which we discuss the content of this course. There is some variation in how linear algebra is presented, just as there is variation in linear optimization presentations. For anyone requiring a quick review of linear algebra or an approach to linear algebra is that this geometry forward, I highly recommend Dr. David Austin's *Understanding Linear Algebra* https://understandinglinearalgebra.org/home.html. This book was motivated by very similar thoughts to the ones I had, and are thereby governed by similar philosophies.

- *This book is presented as a collection of explorations and activities meant to be done by the students in class.* The initial explorations and activities are generally more motivational, followed by activities meant to rigorously crystalize the intuition students develop throughout a lesson. Later activities reinforce these ideas or place them in the broader context of the course.

  These activities are meant to be done in groups by the students, giving them the opportunity to discuss their ideas, let their intuition guide them, and develop their own reasoning skills. I recommend groups of about 4-6, but this is not set in stone, and it should be possible for someone to work on their own, or even self-study with this text. After discussing within their groups, the class comes together to discuss as a single entity.

For the student: I recommend actively engaging in the activities with your peers. We learn best by active participation rather than passive observation. Confronting these ideas, even and especially when you're unsure what is happening is how our understanding grows and deepens. You may be surprised what ideas you come up with, and how much progress you and your classmates can make on your own!

For the instructor: I recommend eavesdropping on group conversations to ensure active participation and equitable practices within groups. It may be necessary to nudge students toward or away from a direction, or to highlight specific things they, or another group said, but as much as possible, try to place agency with the students' hands. Students are far better motivated when they take ownership of their own learning.

It is highly recommended that between classes, students review the activities done in class, and rewrite them to be more cohesive. Frequently, activities will be broken down into a sequence of parts or tasks, each of which represents grabbing a new handhold on the cliff or adjusting the harness to prevent falling (hopefully I am using this metaphor correctly, I am not a rock climber).

Another thing to keep in mind is timing. One challenge of inquiry learning is the amount of time needed to cover content appropriately. By taking a more active role in student discussions, one can help accelerate the process without overtaking the course. However, I recommend that one does so sparingly.

The end of each chapter presents a few exercises that are applications of the material learned, or extensions of those concepts. There is a mix of computational, applied and theoretical problems, and some proof writing will be expected.

- *Interactive technological tools are deployed throughout the text.* In the html version of this book, there will be numerous Doenet activities and Sage cells embedded throughout, there is also embedded Desmos for 3d visualizations. Print and pdf versions of this text will contain QR codes linking to these activities. The use of technology can greatly enhance geometric intuition, or eliminate tedium from computations. Hand computations are a part of the class, and students will be expected to be able to explain and perform them. However, the focus of this text is ever on the concepts of this course, and modeling a problem is almost always a more important skill than computing a solution.

  There are also practical elements to computing solution by code: in class where time is a premium, executing an algorithm with potentially dozens of steps is often a poor use of that limited resource, and in practice, students who go into industry would be working on problems of tremendous scope, far beyond what could be done by hand. Even for students who pursue theoretical mathematics, the intuition and proofs are a far better focus than hand computation.

- *As always, we anchor everything in this course with geometric reasoning, and use this reasoning to bolster the algebraic aspects of this material.* Much like linear algebra, students who find themselves confused or lost in the midst of algebraic weeds, should retreat and consider the situation from a geometric point of view. It can be natural for practical linear optimization problems to take place in hundreds, maybe thousands of dimensions, but two and three dimensions gives us all the intuition we need.

- *I chose presentations and conventions which best support student intuition.* Frustratingly, just about every linear optimization text seems to have their own convention and notation when it comes to recording and presenting data. This book follows Dr. James Strayer's *Linear Programming and it's Applications* as being, in my opinion, the most natural, intuitive, and compact of the innumerable variations.

Linear Optimization: A Geometric Inquiry Course was a labor of love. I hope that this book serves your needs as an instructor, student, or curious scholar. This book can be used as a stand-alone text or supplement another text, with some adjustment. The content of this course is licensed by Creative Commons, and so please feel free to create variations for this material that best suits your needs.

A note on the print version

This book aims to develop readers' ability to reason about linear optimization concepts, and to facilitate thinking and intuition. Several technological tools are deployed to this effect, to enhance geometric reasoning, bypass tedious arithmetic computations, or both. These include embedded Sage cells with executable code, Doenet activities, and Desmos graphs.

Sage is introduced as a platform for performing many computations since it is freely available and its syntax mirrors common mathematical notation. Print readers may access Sage online through https://sagecell.sagemath.org/.

Throughout the book, Sage cells appear in various places to encourage readers to use Sage to complete some relevant computation. In the print version, these may appear with some pre-populated code, such as the one below, that you will want to copy into an online Sage cell.

```
A = matrix([[1,2], [2,1]])
```

Some activities call for the students to determine the appropriate parameters which would allow Sage to solve the question at hand. In these cases, `FIXME` or similar place holders are placed within the code, and meant to be edited out in favor of the correct entries

```
A = matrix(FIXME)
rref(A)
```

Empty cells appear as shown below and are included to indicate part of an exercise or activity that is meant to be completed in Sage.

```

```

For other interactive components to activities, the pdf and print versions of this text will contain links and QR codes to a standalone version of the activity, so that readers with an appropriate device may follow along the flow of the course.

In this section, I provide two additional such interactives. One is a Desmos linear optimization visualizer and solver for two dimensional problems:

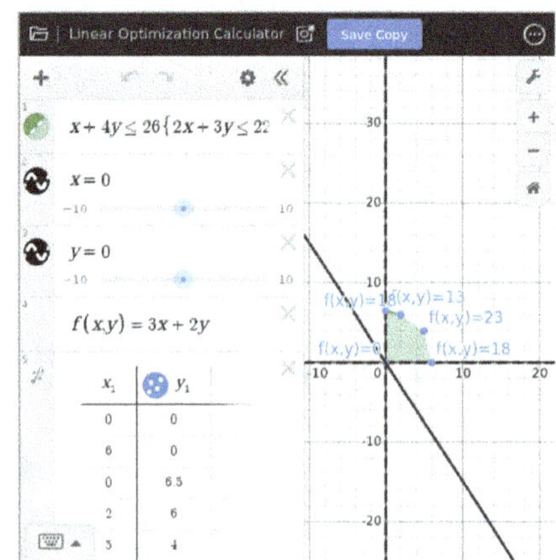

The other is a Desmos 3d version of this interactive for three dimensions.

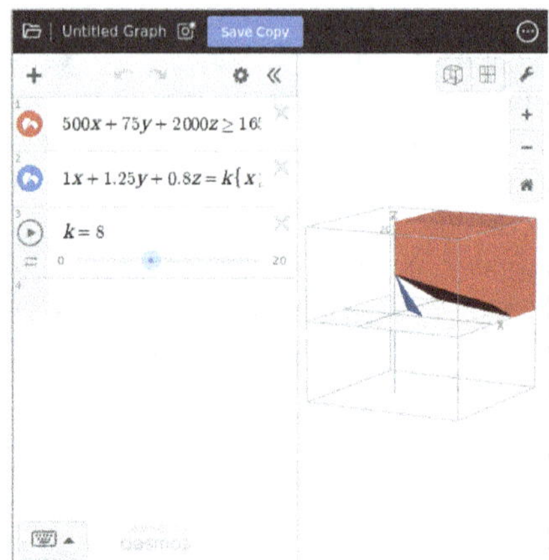

These tools can help you visualize and solve small dimensional problems and can be very helpful for beginners.

There is also a handy simplex pivoting tool provided in Appendix B to simplify computations for higher dimension problems.

A note on inquiry learning

The four pillars of Inquiry Learning are the cornerstone around which this text was built. For those who wish to learn now, I highly reccomend Dr. Dana Ernst's blog on this subject https://danaernst.com/resources/inquiry-based-learning/.

Some care must be taken to ensure that the class flows in a productive way, and that students are equitably engaged, and are given agency to take ownership of their own learning. Some judicious instructor intervention may occasionally be necessary, but this text facilitates a student centered, deep and engaging learning experience.

While it may be possible for a student to work through this text on their own, it is designed to facilitate discussion between students, and between the class and the instructor. There is truly where this material come to life.

I highly recommend, if possible, projecting the html version of this text on a screen, and for students to use whatever computing devices they have available to access this text, use the provided tools, and collaborate on working through this material.

Best wishes to everyone reading this text on your grand adventure through linear optimization!

Contents

Appendices

Back Matter

Chapter 1

Geometric Linear Optimization

The purpose of this chapter is to establish the geometric intuition that undergird the approach we take to this material throughout the course. In Section 1.1 we revisit some ideas from linear algebra and explore the connection between algebraic operations and geometric interpretations. This is *not* meant to serve as a review of linear algebra prerequisites, but as a warm up that centers the geometric intuition (A brief review of a few linear algebra topics may be found in Section A.1). In Section 1.2 we give some examples of problems we wish to address in this course and show how we may interpret them geometrically. In Section 1.3 we describe the geometric notions which will be most useful to us throughout the course, and give formal and intuitive descriptions of these ideas.

1.1 A Brief Geometric Review of Linear Algebra

In this introductory section, we do not begin linear optimization. Instead, we recall a few concepts from linear algebra, and examine them through a geometric lense, setting the stage for our mindset going forward. The activities in this section are not strictly necessary to work through the rest of this text.

Activity 1.1.1 Let
$$A = \begin{bmatrix} 1 & a_2 & a_3 & \cdots & a_m \end{bmatrix}.$$

What is $\dim(\text{null}(A))$?

Activity 1.1.2 Consider the augmented matrix
$$M = \begin{bmatrix} a_{11} & a_{12} & \cdots & a_{1m} & b_1 \\ a_{21} & a_{22} & \cdots & a_{2m} & b_2 \\ \vdots & \vdots & \ddots & \vdots & \vdots \\ a_{n1} & a_{n2} & \cdots & a_{nm} & b_n \end{bmatrix} = (A|b).$$

(a) Given a fixed m, what is a necessary condition for the values of n so that the system of equations encoded by M has a unique solution?

(b) What does this mean geometrically?

(c) If the rows of A are independent, and $n < m$, then what is the dimension of the solution space of M?

Activity 1.1.3 Consider the matrix

$$
M = \begin{bmatrix}
a_{11} & a_{12} & \cdots & a_{1m} \\
a_{21} & a_{22} & \cdots & a_{2m} \\
\vdots & \vdots & \ddots & \vdots \\
a_{n1} & a_{n2} & \cdots & a_{nm}
\end{bmatrix}.
$$

(a) Describe necessary and sufficient conditions for the columns to be linearly independent.

(b) Describe necessary and sufficient conditions for the columns to be a spanning set.

(c) Describe necessary and sufficient conditions for the columns to be a basis for $\mathbb{R}^m$.

Here, we recall the geometric interpretation of the determinant.

Observation 1.1.4 Recall that each $n \times m$ matrix may be thought of as a linear transformation from $\mathbb{R}^m \to \mathbb{R}^n$. When $m = n$ we may define the **determinant** of the transformation. The determinant has numerous algebraic properties one learns about in a linear algebra course, but it also has a geometric interpretation. Roughly speaking, the determinant measures how the transformation changes the unit n dimensional cube, with the magnitude of the determinant measuring the n-dimensional volume of the cube after transformation, and the sign measuring whether or not the orientation of the cube is preserved or reversed.

Some resources for linear algebra define the determinant algebraically, then prove that it has special geometric properties. In many ways this is a natural approach to introduce the subject to students whose background is primarily algebraic. However, in my opinion, this is backwards. It makes far more sense to approach the determinant geometrically first: there is a property of transformations we want to measure, we call this quantity the determinant, it happens to have cool algebraic properties.

Activity 1.1.5 Consider the square matrix

$$
A = \begin{bmatrix}
a_{11} & a_{12} & \cdots & a_{1n} \\
a_{21} & a_{22} & \cdots & a_{2n} \\
\vdots & \vdots & \ddots & \vdots \\
a_{n1} & a_{n2} & \cdots & a_{nn}
\end{bmatrix}.
$$

(a) Explain why for any constant c and $1 \leq j \leq n-1$

$$
\det(A) = \det\left(\begin{bmatrix}
a_{11} & a_{12} & \cdots & a_{1n} \\
a_{21} & a_{22} & \cdots & a_{2n} \\
\vdots & \vdots & \ddots & \vdots \\
a_{n1} + ca_{j1} & a_{n2} + ca_{j2} & \cdots & a_{nn} + ca_{jn}
\end{bmatrix}\right)
$$

geometrically, i.e. without cofactor expansion.

(b) Explain why for any constant c and $1 \leq j \leq n$

$$\det\left(\begin{bmatrix} a_{11} & a_{12} & \cdots & a_{1n} \\ a_{21} & a_{22} & \cdots & a_{2n} \\ \vdots & \vdots & \ddots & \vdots \\ ca_{n1} & ca_{n2} & \cdots & ca_{nn} \end{bmatrix}\right) = c\det(A)$$

geometrically.

(c) Explain why for any constant c and $1 \leq j \leq n$

$$\det\left(\begin{bmatrix} a_{11} & a_{12} & \cdots & a_{1(n-1)} & 0 \\ a_{21} & a_{22} & \cdots & a_{2(n-1)} & 0 \\ \vdots & \vdots & \ddots & \vdots & \vdots \\ a_{(n-1)1} & a_{(n-1)2} & \cdots & a_{(n-1)(n-1)} & 0 \\ 0 & 0 & \cdots & 0 & 1 \end{bmatrix}\right)$$

$$= \det\left(\begin{bmatrix} a_{11} & a_{12} & \cdots & a_{1(n-1)} \\ a_{21} & a_{22} & \cdots & a_{2(n-1)} \\ \vdots & \vdots & \ddots & \vdots \\ a_{(n-1)1} & a_{(n-1)2} & \cdots & a_{(n-1)(n-1)} \end{bmatrix}\right)$$

geometrically.

Activity 1.1.6 Michael Atiyah (1929 - 2019), mathematician and Field's medalist (1966), once said:

"Algebra is the offer made by the devil to the mathematician. The devil says: 'I will give you this powerful machine, it will answer any question you like. All you need to do is give me your soul: give up geometry and you will have this marvelous machine.'"

What do you suppose Dr. Atiyah meant by this quote? What does it mean to you? How might this sentiment have impacted your mathematical journey or education?

1.2 Initial Examples

Here we begin with some initial examples motivating the sort of problems we will study. We define the central problems around which the course will revolve.

Activity 1.2.1 A sculptor and a painter work together to produce pieces of art, vases and figurines. The vases takes the sculptor 1 hour to make and the painter 2 hours to paint. The figurine takes the sculptor 2 hours to make and the painter 1 hour to paint. The sculptor has 35 hours a week to work and the painter has 40 hours a week to work.

Let x_1 denote the number of vases produced and x_2 denote the number of figurines produced.

(a) Write an inequality terms of x_1 and x_2 that represent the constraints on the time of the sculptor.

(b) Write an inequality terms of x_1 and x_2 that represent the constraints on the time of the painter.

(c) What other inequalities involving x_1 and x_2 would be sensible to impose?

(d) Treating x_1 as x and x_2 as y, sketch the region on the Cartesian plane satisfying all the above inequalities. We will refer to this as the **feasible region**.

(e) Pick some points within this feasible region, what do they represent in terms of vases and figurines? How much revenue is generated? What causes the revenue to increase or decrease?

(f) Suppose the vases sold for \$100 and the figurines sold for \$120. Without reading ahead, what would or could you do to solve this problem? What kind of things would you need to consider?

(For those of you who had Calculus and especially Multivariable Calculus, what ideas from those course might you employ? What are some limitations of these ideas?)

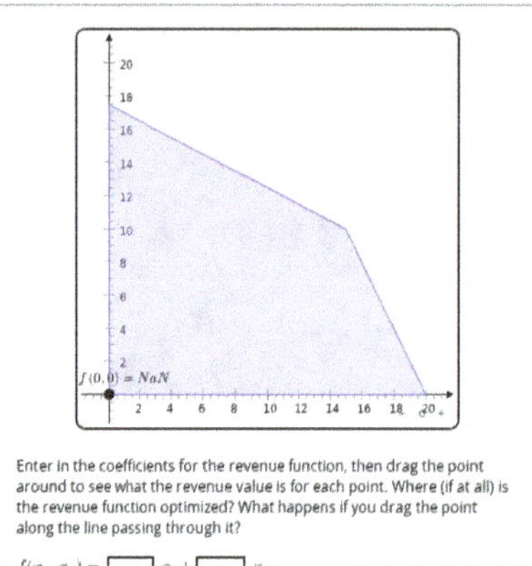

Enter in the coefficients for the revenue function, then drag the point around to see what the revenue value is for each point. Where (if at all) is the revenue function optimized? What happens if you drag the point along the line passing through it?

Standalone
Embed

(g) If there was a surge in demand for vases, and they started selling for \$1000, how would that change your approach and the solution?

Activity 1.2.2 Carlos is planning to shop for a meal. Mini bell peppers are \$1.00 (simplified) a pound, Chicken is \$1.25 a pound and Spaghetti with sauce is \$0.80 a cup.

A pound of Bell Peppers has 500 units of vitamin A, 30 calories and 10 units of calcium. A pound of Chicken has 75 units of vitamin A, 280 calories and 22 units of calcium. A cup of Spaghetti with sauce has 2000 units of Vitamin A, 240 Calories and 52 units of Calcium.

Carlos needs a minimum of 640 calories, 1650 units of Vitamin A and 500 units of Calcium.

(a) Let x_1 denote pounds of bell pepper, x_2 denote pounds of chicken and x_3 denote cups of spaghetti with sauce. Find three inequalities in terms of the x_i for how much of each food Carlos should eat to meet his minimum dietary requirements.

(b) How might we solve this problem? How is it different from Activity 1.2.1?

(c) This seems like a wildly over simplistic dietary problem, because it is. How might we complicate it for more realism?

Definition 1.2.3 The **Canonical Maximization Linear Optimization Problem** is the problem:

$$\textbf{Maximize: } f(\mathbf{x}) = c_1 x_1 + c_2 x_2 + \cdots c_m x_m - d = \left(\sum_{j=1}^{m} c_j x_j \right) - d$$

$$\textbf{subject to: } a_{11} x_1 + a_{12} x_2 + \cdots a_{1m} x_m \leq b_1$$
$$a_{21} x_1 + a_{22} x_2 + \cdots a_{2m} x_m \leq b_2$$
$$\vdots$$
$$a_{n1} x_1 + a_{n2} x_2 + \cdots a_{nm} x_m \leq b_n$$
$$x_1, x_2, \ldots, x_m \geq 0$$

where $a_{ij}, b_i, c_j, d \in \mathbb{R}$.

The **Canonical Minimization Linear Optimization Problem** is the problem:

$$\textbf{Minimize: } g(\mathbf{x}) = c_1 x_1 + c_2 x_2 + \cdots c_n x_m - d = \left(\sum_{j=1}^{m} c_j x_j \right) - d$$

$$\textbf{subject to: } a_{11} x_1 + a_{12} x_2 + \cdots a_{1m} x_m \geq b_1$$
$$a_{21} x_1 + a_{22} x_2 + \cdots a_{2m} x_m \geq b_2$$
$$\vdots$$
$$a_{n1} x_1 + a_{n2} x_2 + \cdots a_{nm} x_m \geq b_n$$
$$x_1, x_2, \ldots, x_m \geq 0$$

where $a_{ij}, b_i, c_j, d \in \mathbb{R}$. $\Diamond$

Note 1.2.4 The term **canonical** in this context refers to the fact that for both of the above problems, the x_i are nonnegative, and each bound is an inequality.

(In Chapter 3, we discuss *noncanonical* linear optimization problems, where these conditions may fail.)

Definition 1.2.5 f, g above are called **objective functions**. Any point $\mathbf{x} = (cx_1, x_2, \ldots, x_m)$ satisfying either of the above set of inequalities are called **feasible solutions**. Any feasible solution $\mathbf{x}$ which maximizes (minimizes) the objective function is called an **optimal solution**. $\Diamond$

Activity 1.2.6

(a) Given the canonical minimization problem:

$$\textbf{Minimize: } g(x_1, x_2) = 5x_1 - 3x_2$$
$$\textbf{subject to: } 4x_1 + x_2 \geq 10$$
$$3x_1 + 12x_2 \geq 12$$
$$x_1, x_2, \geq 0.$$

How might we convert this to a canonical maximization problem?

(b) How might we in general convert a minimization problem to a maximization problem?

1.3 Polyhedral Convextiy

In this section, we establish the fundamental geometric notions which underlie
our work.

Activity 1.3.1 In $\mathbb{R}^3$, describe geometrically what the following represent.

(a) $2x - y + 3z = 4$

(b) $2x - y + 3z = 0$

(c) $2x - y + 3z = -4$

(d) $2x - y + 3z \leq 3$

(e) $2x - y + 3z \geq -1$

Activity 1.3.2 Let $\mathbf{p} = (-3, 1), \mathbf{q} = (2, -5)$. Let $\mathbf{x} = t \cdot \mathbf{p} + (1 - t) \cdot \mathbf{q}$ for some
$0 \leq t \leq 1$.

(a) What is $\mathbf{x}$ when $t = 0.2$?

(b) What is $\mathbf{x}$ when $t = 0.5$?

(c) What is $\mathbf{x}$ when $t = 0$?

(d) What is $\mathbf{x}$ when $t = 1$?

(e) Describe the set of points $\{t \cdot \mathbf{p} + (1 - t) \cdot \mathbf{q} : t \in [0, 1]\}$.

(f) Let $\mathbf{p} = (2, 1, 0), \mathbf{q} = (0, -3, 1)$. Describe the set of points $\{t \cdot \mathbf{p} + (1 - t) \cdot \mathbf{q} : t \in [0, 1]\}$.

Definition 1.3.3 Given $\mathbf{p}, \mathbf{q} \in \mathbb{R}^m$, we call the set $\{t \cdot \mathbf{p} + (1 - t) \cdot \mathbf{q} : t \in [0, 1]\}$
the **line segment** between $\mathbf{p}$ and $\mathbf{q}$. $\diamond$

Definition 1.3.4 Given constants $a_1, \ldots, a_m, b$ **hyperplane** in $\mathbb{R}^m$ is a set
of the form $H = \{\mathbf{x} \in \mathbb{R}^m : a_1 x_1 + a_2 x_2 + \cdots a_m x_m = b\}$. If $m = 3$ we usually
call this a **plane**.

$$\text{We call the vector } \begin{bmatrix} a_1 \\ \vdots \\ a_m \end{bmatrix} \text{ the } \textbf{normal vector} \text{ to } H. \qquad \diamond$$

Activity 1.3.5 Consider the desmos interactive:

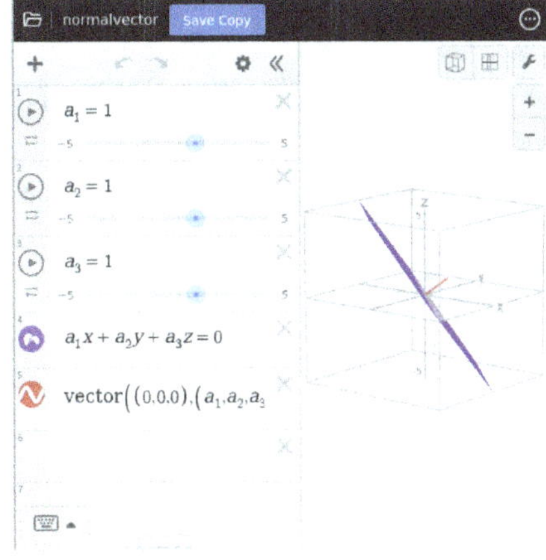

Standalone

Adjust the different a_i sliders to different values. What do we notice about the plane $a_1 x + a_2 y + a_3 z = 0$ and the vector $\begin{bmatrix} a_1 \\ a_2 \\ a_3 \end{bmatrix}$?

Definition 1.3.6 Let $S \subseteq \mathbb{R}^m$, we say that S is **convex** if given any $\mathbf{p}, \mathbf{q} \in S$, S also contains the line segment between $\mathbf{p}, \mathbf{q}$. ◊

Activity 1.3.7 For each of the following subsets of $\mathbb{R}^2$, sketch the region, decide if it is convex or not.

(a) $\{(x, y) : 2x + 3y \le 4\}$.

(b) $\{(x, y) : y = 2 \text{ and } x < 4\}$.

(c) $\{(r, \theta) : 0 \le r \le 1, \theta \in [0, \pi/4]\}$ (in polar coordinates).

(d) $\{(r, \theta) : 0 \le r \le 1, \theta \in [\pi/4, 2\pi]\}$ (in polar coordinates).

(e) $\mathbb{R}^2$.

(f) $\emptyset$.

Activity 1.3.8 Let $S \subseteq \mathbb{R}^m$ be defined by $S := \{\mathbf{x} \in \mathbb{R}^m : a_1 x_1 + a_2 x_2 + \cdots a_m x_m \le b\}$ for some $a_i, b \in \mathbb{R}$ (i.e. S is a half-space of $\mathbb{R}^m$).

(a) Let $f(\mathbf{x}) : \mathbb{R}^m \to \mathbb{R}, (x_1, x_2, \ldots, x_m) \mapsto a_1 x_1 + a_2 x_2 + \cdots a_m x_m$. Explain why $f(c\mathbf{x}) = cf(\mathbf{x})$ for any $c \in \mathbb{R}$.

(b) Show that for any $\mathbf{x} = (x_1, x_2, \ldots, x_n), \mathbf{y} = (y_1, y_2, \ldots, y_n)$, that $f(\mathbf{x} + \mathbf{y}) = f(\mathbf{x}) + f(\mathbf{y})$.

(c) Let $\mathbf{x}, \mathbf{y} \in S$, that is, there are k_x, k_y such that $f(\mathbf{x}) = a_1 x_1 + a_2 x_2 + \cdots a_n x_n = k_x, f(\mathbf{y}) = a_1 y_1 + a_2 y_2 + \cdots a_n y_n = k_y$ and $k_x, k_y \le b$. Show that $f(t\mathbf{x} + (1 - t)\mathbf{y}) \le b$ for $t \in [0, 1]$.

(d) Conclude that S is convex.

Activity 1.3.9 Let S_1, S_2 be convex sets.

(a) Show that $S_1 \cap S_2$ is convex.

   Hint. Let $P, Q \in S_1 \cap S_2$. Why is the line segment between them contained in S_1? S_2?

(b) Sketch an induction argument to show that if S_i is convex, $\bigcap_{i=1}^{n} S_i$ is convex.

(c) Prove or find a counterexample to the following statement: If S_1, S_2 are convex sets, then $S_1 \cup S_2$ is convex.

Activity 1.3.10 Prove that the feasible region of a canonical linear optimization problem is convex.

Definition 1.3.11 A convex set S that is equal to a finite intersection of half-spaces (defined by either strict or nonstrict inequalities) is **polyhedral convex**. ◊

Example 1.3.12 The feasible region for Activity 1.2.1 is:

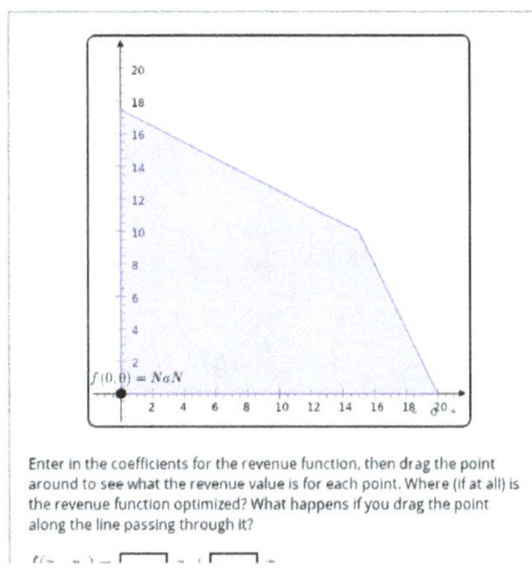

Enter in the coefficients for the revenue function, then drag the point around to see what the revenue value is for each point. Where (if at all) is the revenue function optimized? What happens if you drag the point along the line passing through it?

Example 1.3.13 The feasible region for Activity 1.2.2 is:

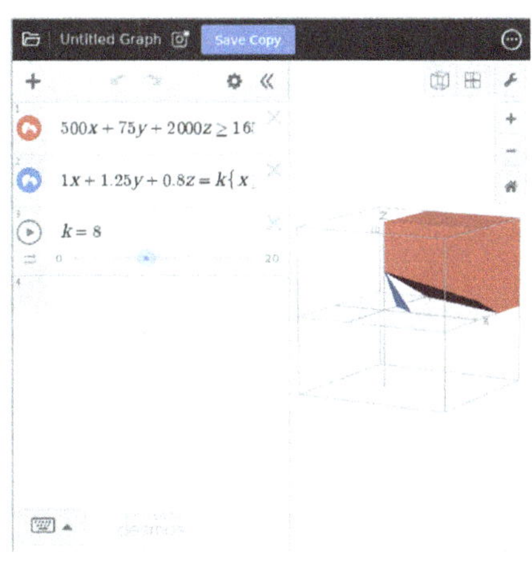

Definition 1.3.14 Given $\mathbf{x} \in \mathbb{R}^n$ we define the **norm** of $\mathbf{x}$ to be

$$\|\mathbf{x}\| = \sqrt{x_1^2 + x_2^2 + \cdots x_n^2}.$$

Definition 1.3.15 Given $\mathbf{x} \in \mathbb{R}^n$ we define the **closed ball** of radius r centered at $\mathbf{x}$ to be

$$\bar{B}(\mathbf{x}, r) := \{\mathbf{y} : \|\mathbf{x} - \mathbf{y}\| \leq r\}.$$

The **open ball** centered at $\mathbf{x}$ with radius r is similarly defined. What do you think it is?

Activity 1.3.16 Describe $\bar{B}(\mathbf{0}, r)$ for $\mathbb{R}, \mathbb{R}^2, \mathbb{R}^3$.

Definition 1.3.17 A set S is bounded if there is $r \geq 0$ such that $S \subseteq \bar{B}(\mathbf{0}, r)$.

Activity 1.3.18 Which of the problems described in Activity 1.2.1 and Activity 1.2.2 have bounded feasible regions?

Activity 1.3.19 Let $S \subseteq \mathbb{R}^2$ be the feasible region of of a canonical maximization linear optimization problem, let $f(x_1, x_2) = 2x_2 + 3x_2$ be the objective function.

(a) Consider the point $(1, 1)$. Which direction would increase the value of f

the most? The least? Keep f the same?

(Recall the properties of the dot product.)

(b) Let $\mathbf{x} \in S$ such that there is a $r > 0$ so that $B(\mathbf{x}, r) \subseteq S$. Explain why $\mathbf{x}$ cannot be a maximizer of f.

(c) On the other hand, suppose $\mathbf{x}^* \in S$ *is* a maximizer of f, what must be true about $\mathbf{x}^*$?

(d) Consider the canonical maximization linear optimization problem:

$$\textbf{Maximize: } f(x, y) = 3x + 2y$$
$$\textbf{subject to: } x + y \leq 8$$
$$2x + y \leq 10$$
$$x, y \geq 0.$$

How do the statements you've made above apply here? Where *is* f maximized? Is it consistent with what you said before?

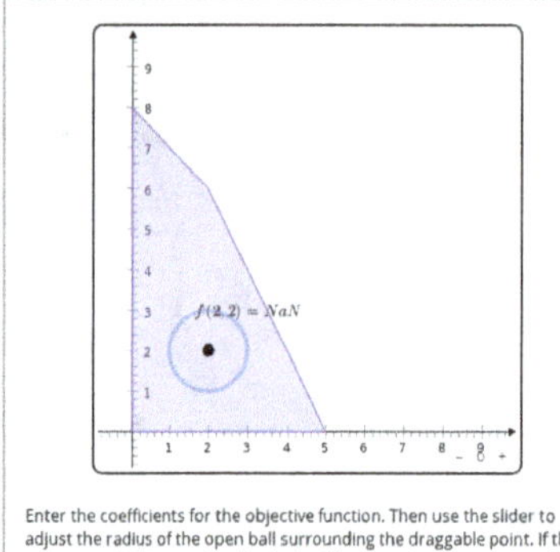

Enter the coefficients for the objective function. Then use the slider to adjust the radius of the open ball surrounding the draggable point. If the entire open ball is contained in the feasible region for a given point, why can't this point be an optimizer? Can we move the point to a different feasible point with higher objective value?

Standalone
Embed

Definition 1.3.20 Let $S \subseteq \mathbb{R}^m$ be a convex set. We say $\mathbf{e} \in S$ is an **extreme point** of S if there are no distinct $\mathbf{x}, \mathbf{y} \in S$ so that $\mathbf{e} \in \{t \cdot \mathbf{x} + (1-t) \cdot \mathbf{y} : t \in (0, 1)\}$.

In other words, $\mathbf{e}$ does not lie on any nontrivial line segment contained in S. ◊

Activity 1.3.21 For each of the following convex sets, find its extreme points (if any).

(a) The feasible region of the problem in Activity 1.3.19.

(b) $\{(x, y) : \|(x, y)\| \leq 1\} \subseteq \mathbb{R}^2$.

(c) $\{(x, y) : y \geq 0\} \subseteq \mathbb{R}^2$.

We present the following theorems without proof. At least one of these should seem familiar from calculus.

Theorem 1.3.22 Extreme Value Theorem. *If $S \subseteq \mathbb{R}^m$ is a closed and bounded region, and $f : S \to \mathbb{R}$ is a continuous function. Then there are $\mathbf{x}_1, \mathbf{x}_2 \in S$ such that $f(\mathbf{x}_1) \geq f(\mathbf{x})$ and $f(\mathbf{x}_2) \leq f(\mathbf{x})$ for every $\mathbf{x} \in S$.*

(We will assume without proof that linear functions $f : \mathbb{R}^m \to \mathbb{R}$ are continuous, and that the feasible region of a linear optimization problem is always closed.)

Theorem 1.3.23 *If S is the feasible region of a canonical problem and is bounded, then S contains an optimal solution which is an extreme point of S.*

Theorem 1.3.24 *If S is the feasible region of a canonical problem and is unbounded:*

1. *If the problem is a maximization problem and if there is an M so that $f(s) \leq M$ for all $s \in S$, then S contains an optimal solution which is an extreme point of S.*

2. *If the problem is a minimization problem and if there is an M so that $g(s) \geq M$ for all $s \in S$, then S contains an optimal solution which is an extreme point of S.*

Activity 1.3.25

(a) Let $f(x, y) = x + y$ be the objective function for a canonical maximization problem, subject to:

$$x + 4y \leq 26$$
$$2x + 2y \leq 16$$
$$4x + y \leq 24$$
$$x, y \geq 0.$$

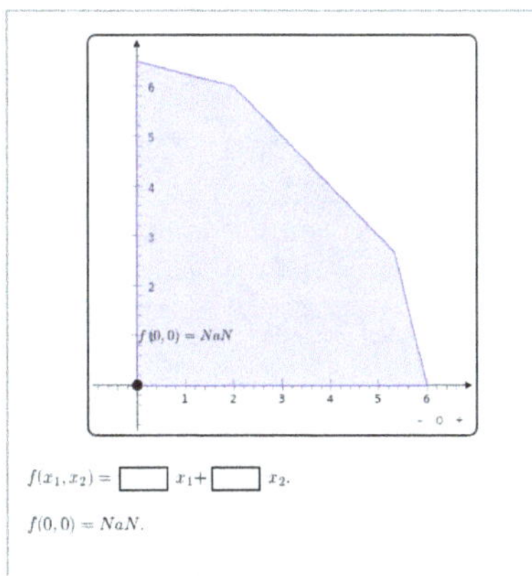

Standalone
Embed

Find a maximal solution that is not a corner point. Why doesn't this contradict Theorem 1.3.23?

(b) Let $\mathbf{x}_1, \mathbf{x}_2 \in S$ be optimal solutions of a canonical linear optimization problem, giving optimal solution $f(\mathbf{x}_1) = f^* = f(\mathbf{x}_2)$. Show that for any $t \in [0, 1]$, $f(t\mathbf{x}_1 + (1 - t)\mathbf{x}_2) = f^*$ and that $t\mathbf{x}_1 + (1 - t)\mathbf{x}_2 \in S$.

(c) Let $f(x, y) = x$ be the objective function for a canonical maximization problem. Find a set of constraints so that the feasible region is unbounded but there is a maximal solution. Why doesn't this contradict Theorem 1.3.24?

Activity 1.3.26 So, we see that the hunt for optimal solutions boils down to a hunt for extreme points.

(a) In $\mathbb{R}^2$, how many lines are needed to intersect so their intersection is a unique point?

(b) In $\mathbb{R}^3$, how many planes are needed to intersect so their intersection is a unique point?

(c) In $\mathbb{R}^m$, how many $m - 1$ dimensional hyperplanes are needed to intersect so their intersection is a unique point?

(d) Suppose a canonical linear optimization problem in $\mathbb{R}^m$ is bounded by the usual m hyperplanes corresponding to $x_i \geq 0$ as well as k additional hyperplanes. How many potential points of intersection could there be?

A. n

B. $\dbinom{m}{k}$

C. $\dbinom{m+k}{k}$

D. $\dbinom{k}{m}$

E. $\dbinom{m+k}{n}$

(e) So for a canonical linear optimization problem in $\mathbb{R}^{10}$ bounded by an additional 15 hyperplanes, how many potential extreme points are there?

1.4 Summary of Chapter 1

We introduced some basic examples of linear optimization problems in Activity 1.2.1 and Activity 1.2.2. In these examples, we were able to graphically analyze and show that they reach optimality, and that these optimal solutions occur on the boundary of the feasible region. This is an idea which is critical to the development of our theory.

To show that such a result may generalize, we introduce the notion of **convexity** Definition 1.3.6. A convex set is a set which contains every line segment between points in the set. We then show that a half-space is convex, that an intersection of halfspaces is convex, and thus the feasible region of a linear optimization problem is always convex.

Note then that for a linear function $f(x_1, x_2, \ldots x_m) = \sum_{j=1}^{m} c_j x_j$, that for any output of f, k, the equation

$$\sum_{j=1}^{m} c_j x_j = k$$

forms a plane in $\mathbb{R}^m$, and that for any $\mathbf{x}'$ lying on this plane, $f(\mathbf{x}') = k$ as well. We may increase the value of k by moving in one of two directions orthogonal to this plane, until we reach the boundary. But even then, we can still increase the value of k by moving along the boundary until we either fall on a subset of the boundary on which f is constant, or a corner point also called an **extreme point** Definition 1.3.20.

Thus, if a linear optimization problem admits a maximum solution, it must do so at some extreme point. Each extreme point is an intersection of hyperplanes, but in $\mathbb{R}^m$ with k additional bounding hyperplanes, there could be $\binom{m+k}{m}$ intersections, not all of whom are feasible, and not all of whom are optimal. There is also no guarantee that a linear optimization problem admits an optimal solution at all. This motivates us to find a more systematic approach to obtaining optimal solutions.

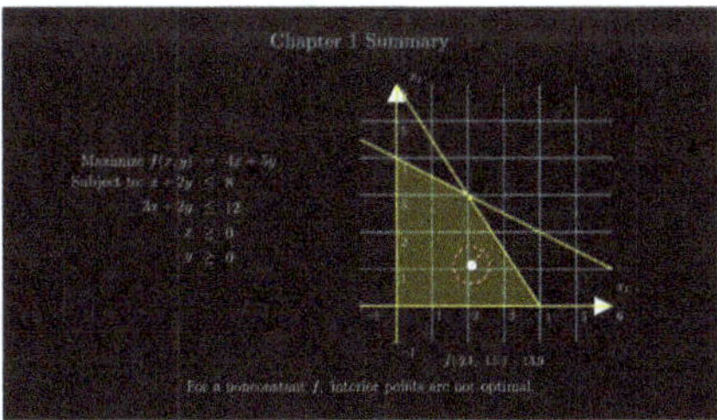

Standalone

Figure 1.4.1 A summary of Chapter 1.

1.5 Problems for Chapter 1

Exercises

1. Draw and shade appropriate regions in $\mathbb{R}^2$ as described below, where $x, y \geq 0$.

   (a) A bounded polyhedral convex subset.

   (b) An unbounded polyhedral convex subset.

   (c) A bounded nonconvex subset

   (d) An unbounded nonconvex subset.

   (e) A convex subset that is not a polyhedral convex subset.

   (f) A convex subset having no extreme points.

   (g) A polyhedral convex subset having no extreme points.

   (h) A bounded polyhedral convex subset having exactly one extreme point.

   (i) An unbounded polyhedral convex subset having exactly one extreme point.

   (j) An bounded convex subset having infinitely many extreme points.

   (k) An unbounded convex subset having infinitely many extreme points.

2. Convert each of the linear optimization problems below to canonical form as in Definition 1.2.3.

   (a)

$$\textbf{Maximize: } f(x,y) = 3x - y$$
$$\textbf{subject to: } 15 - 4x \geq y$$
$$2x + 3y \leq y + 12$$
$$x, y \geq 0.$$

   (b)

$$\textbf{Minimize: } g(x,y) = -4x - 2y$$
$$\textbf{subject to: } 4x - 7y \leq -2$$
$$5x + 3y \leq 21$$
$$x, y \geq 0.$$

   (c)

$$\textbf{Maximize: } h(x,y,z) = x + 3y - 3z$$
$$\textbf{subject to: } 3 - 4x + z \geq 5y$$
$$1 \leq 2y + z \leq 9$$
$$0 \leq x \leq 8$$
$$y, z \geq 0.$$

(d)

$$\text{Minimize: } k(x, y, z) = x - 2y - z$$
$$\text{subject to: } 10x + 5y + 2z \leq 1000$$
$$2y + 4z \leq 800$$
$$x, y, z \geq 0.$$

(e) A drive-in sells homemade hot dogs and hamburgers. The hot dogs take 3/8 cup of flour and 2.5 oz of beef to make. A hamburger bun takes 1/4 cups of flour and 5 oz of beef. Suppose the drive-in has 20 cups of flour and 200 oz of beef on hand. If hot dogs sell for $4 and hamburgers for $6, how much of each should they make to maximize revenue?

(f) A rancher has a herd to feed which requires 54, 48, and 72 units of the nutritional elements A, B, and C, respectively, per day. Feed 1 costs 10 cents a pound and contains 8, 4 and 3 units of elements A, B, C respectively. Feed 2 costs 8 cents a pound and contains 2, 4 and 6 units of elements A, B, C respectively. How much of each feed should the rancher purchase to cover the herds nutritional needs while minimizing cost?

(g) A drug company sells three different formulations of vitamin complex and mineral complex. The first formulation consists entirely of vitamin complex and sells for $1 per unit. The second formulation consists of 3/4 of a unit of vitamin complex and 1/4 of a unit of mineral complex and sells for $2 per unit. The third formulation consists of 1/2 of a unit of each of the complexes and sells for $3 per unit. If the company has 100 units of vitamin complex and 75 units of mineral complex available, how many units of each formulation should the company produce so as to maximize sales revenue?

(h)

$$\text{Maximize: } f(x, y) = 5x + 6y$$
$$\text{subject to: } 2x + 3y \geq 12$$
$$-3x + 2y \leq 14$$
$$x + y \leq 12$$
$$x, y \geq 0.$$

3. For each of the following, sketch the feasible region **R**, and find the optimal solution by identifying the extreme points of **R** and evaluating.

(a)

$$\text{Maximize: } f(x, y) = 2x + y$$
$$\text{subject to: } x + 4y \leq 11$$
$$6x + 2y \leq 22$$
$$x, y \geq 0.$$

(b)

$$\text{Minimize: } f(x, y) = 2x + y$$
$$\text{subject to: } x + 4y \geq 11$$

$$6x + 2y \geq 22$$
$$x, y \geq 0.$$

(c) Exercise 1.5.2 **(a)**.

(d) Exercise 1.5.2 **(b)**.

(e) Exercise 1.5.2 **(d)**.

(f) Exercise 1.5.2 **(e)**.

(g) Exercise 1.5.2 **(f)**.

(h) Exercise 1.5.2 **(g)**.

(i) Exercise 1.5.2 **(h)**.

4. Prove that Exercise 1.5.3 **(h)** has infinitely many optimal solutions, two of which lie on extreme points. Identify these points on the plot of the feasible region done in Exercise 1.5.3.

5. Prove that if $\mathbf{x}, \mathbf{x}'$ are distinct optimal solutions of a canonical linear optimization problem, then all points on the line segment between $\mathbf{x}, \mathbf{x}'$ are also optimal solutions of the problem.

Conclude that a canonical linear optimization problem can have 0, 1 or infinitely many optimal solutions and no other possibilities.

6. Consider the canonical linear optimization problem

> **Maximize:** $f(x, y, z, w) = 3x - 2y + 5z - w$
>
> **subject to:** $x + 2y + 3z + 6w \leq 12$
>
> $$x, y, z, w \geq 0.$$

Find the $\binom{5}{4}$ potential intersections of bounding hyperplanes, determine which are feasible, and which of those are optimal.

7. Consider the canonical linear optimization problem

> **Minimize:** $g(x, y, z, w) = 2x + y + 3z + 4w$
>
> **subject to:** $x + 2y + 2z + 4w \geq 8$
>
> $$4x - y \geq 6$$
> $$x, y, z, w \geq 0.$$

Find the $\binom{6}{4}$ potential intersections of bounding hyperplanes, determine which are feasible, and which of those are optimal. (It is recommended that you use some technological tools to solve for the resulting linear systems.)

```
A=matrix([
[1,0,0,0,0],
[0,1,0,0,0],
[0,0,1,0,0],
[0,0,0,1,0]])
print(A)
print("␣")
print(A.rref())
```

8. Express $\binom{k+4}{4}$ as a polynomial in terms of k. How does this relate to Exercise 1.5.6 and Exercise 1.5.7?

9. Show that

$$\textbf{Maximize: } f(x, y) = x - y$$
$$\textbf{subject to: } x + y \geq 5$$
$$x - 5y \leq 0$$
$$y - 2x \leq 1$$
$$x, y \geq 0.$$

is unbounded. (Hint: sketch the feasible region and consider feasible points on the line $x = 5y$.)

10. Show that

$$\textbf{Minimize: } g(x, y) = 5x + 3y$$
$$\textbf{subject to: } y - 5x \geq 1$$
$$x - 5y \geq 0$$
$$x, y \geq 0.$$

is infeasible.

11. For each of the following, determine whether or not the statement is TRUE or FALSE. If TRUE, provide a proof, if FALSE, provide a counterexample.

(a) If a canonical feasible linear optimization problem is unbounded, then its feasible region is unbounded.

(b) If a canonical feasible linear optimization problem has unbounded feasible region, then it is unbounded.

Chapter 2

The Simplex Algorithm

We saw that in lower dimensions one may use geometric reasoning to find and interpret the optimal solution to a linear optimization problem. However, this intuition easily breaks down in higher dimensions. Moreover, brute force examination of potential corner points proves to be a massive and overwhelming approach to finding optimal solutions. Is there any way we can leverage this intuition built in lower dimensions to develop a systematic way to identify the optimal extreme point?

In this chapter, we develop an algorithm, the Simplex Algorithm, which allows us to solve linear optimization problems. In Section 2.1 we determine the rules for a simplex pivot and interpret this operation computationally, algebraically and geometrically. In Section 2.2 we discover the conditions for an appropriate pivot which preserves feasibility, as well as how to handle basic solutions which are infeasible, and minimization problems. In Section 2.3 we discuss *cycling*, and how to avoid it with Bland's Anticycling Rules. Finally in Section 2.4 we show how the programming language Sage may be used to remove some of the computational tedium of solving these problems.

2.1 Canonical Problems and the Simplex Pivot

In this section, we use our geometric intuition to develop an algebraic operation called a **simplex pivot**, which will be the key operation in solving linear optimization problems. We give both a geometric and algebraic understanding of this operation.

Activity 2.1.1 Consider the canonical maximization problem:

$$\textbf{Maximize: } f(\mathbf{x}) = c_1 x_1 + c_2 x_2 + \cdots c_m x_m - d = \left(\sum_{j=1}^{m} c_j x_j \right) - d$$

$$\textbf{subject to: } a_{11} x_1 + a_{12} x_2 + \cdots a_{1m} x_m \leq b_1$$

$$a_{21} x_1 + a_{22} x_2 + \cdots a_{2m} x_m \leq b_2$$

$$\vdots$$

$$a_{n1} x_1 + a_{n2} x_2 + \cdots a_{nm} x_m \leq b_n$$

$$x_1, x_2, \ldots, x_m \geq 0$$

Note that we may rewrite this as a system of equalities by introducing the slack variables t_i. We will refer to the x_j as **decision variables**.

$$\textbf{Maximize: } f(\mathbf{x}) = c_1 x_1 + c_2 x_2 + \cdots c_m x_m - d = \left(\sum_{j=1}^{m} c_j x_j \right) - d$$

$$\textbf{subject to: } a_{11} x_1 + a_{12} x_2 + \cdots a_{1m} x_m + t_1 = b_1$$
$$a_{21} x_1 + a_{22} x_2 + \cdots a_{2m} x_m + t_2 = b_2$$
$$\vdots$$
$$a_{n1} x_1 + a_{n2} x_2 + \cdots a_{nm} x_m + t_n = b_n$$
$$x_1, x_2, \ldots, x_m \geq 0.$$

(a) What must be true about the slack variables t_i?

    A. $t_i \leq 0$.

    B. $t_i \geq 0$.

    C. $\displaystyle\sum_{i=1}^{m} t_i = 0$.

(b) Which of these *must* be an upper bound for t_i?

    (a) b_i.

    (b) c_i.

    (c) x_i.

    (d) t_i could be unbounded.

Observation 2.1.2 Similarly the canonical minimization problem is:

$$\textbf{Minimize: } g(\mathbf{x}) = c_1 x_1 + c_2 x_2 + \cdots c_m x_m - d = \left(\sum_{j=1}^{m} c_j x_j \right) - d$$

$$\textbf{subject to: } a_{11} x_1 + a_{12} x_2 + \cdots a_{1m} x_m \geq b_1$$
$$a_{21} x_1 + a_{22} x_2 + \cdots a_{2m} x_m \geq b_2$$
$$\vdots$$
$$a_{n1} x_1 + a_{n2} x_2 + \cdots a_{nm} x_m \geq b_n$$
$$x_1, x_2, \ldots, x_m \geq 0$$

where $a_{i,j}, b_i, c_j, d \in \mathbb{R}$. This problem may be rewritten as:

$$\textbf{Minimize: } g(\mathbf{x}) = c_1 x_1 + c_2 x_2 + \cdots c_m x_m - d = \left(\sum_{j=1}^{m} c_j x_j \right) - d$$

$$\textbf{subject to: } a_{11} x_1 + a_{12} x_2 + \cdots a_{1m} x_m - s_1 = b_1$$
$$a_{21} x_1 + a_{22} x_2 + \cdots a_{2m} x_m - s_2 = b_2$$
$$\vdots$$
$$a_{n1} x_1 + a_{n2} x_2 + \cdots a_{nm} x_m - s_n = b_n$$
$$x_1, x_2, \ldots, x_m \geq 0.$$

In this case, we refer to the s_i as the **slack variables**, and the x_j as **decision variables**.

As usual, we focus on maximization for now.

Definition 2.1.3 We can rewrite

$$\textbf{Maximize: } f(\mathbf{x}) = c_1 x_1 + c_2 x_2 + \cdots c_m x_m - d = \left(\sum_{j=1}^{m} c_j x_j \right) - d$$

$$\textbf{subject to: } a_{1,1} x_1 + a_{1,2} x_2 + \cdots a_{1,m} x_m + t_1 = b_1$$
$$a_{21} x_1 + a_{22} x_2 + \cdots a_{2m} x_m + t_2 = b_2$$
$$\vdots$$
$$a_{n1} x_1 + a_{n2} x_2 + \cdots a_{nm} x_m + t_n = b_n$$
$$x_1, x_2, \ldots, x_m \geq 0$$

as

$$\textbf{Maximize: } f(\mathbf{x}) = c_1 x_1 + c_2 x_2 + \cdots c_m x_m - d = \left(\sum_{j=1}^{m} c_j x_j \right) - d$$

$$\textbf{subject to: } a_{11} x_1 + a_{12} x_2 + \cdots a_{1m} x_m - b_1 = -t_1$$
$$a_{21} x_1 + a_{22} x_2 + \cdots a_{2n} x_m - b_2 = -t_2$$
$$\vdots$$
$$a_{n1} x_1 + a_{n2} x_2 + \cdots a_{nm} x_m - b_n = -t_n$$
$$x_1, x_2, \ldots, x_m, t_1, \ldots, t_n \geq 0$$

Which may be recorded by the **Tucker tableau**:

x_1	x_2	$\cdots$	x_m	-1	
a_{11}	a_{12}	$\cdots$	a_{1m}	b_1	$= -t_1$
a_{21}	a_{22}	$\cdots$	a_{2m}	b_2	$= -t_2$
$\vdots$	$\vdots$	$\ddots$	$\vdots$	$\vdots$	$\vdots$
a_{n1}	a_{n2}	$\cdots$	a_{nm}	b_n	$= -t_n$
c_1	c_2	$\cdots$	c_m	d	

The variables at the top are called the **decision variables** or **independent variables**, the variables on the side are the **slack variables** or **basic variables**.

The **basic solution** recorded by a Tucker tableau is the solution where each decision variable has the value zero. $\diamond$

Observation 2.1.4 Note that this tableau records a basic solution $x_1, x_2, \ldots, x_m = 0$. We will further explore what this means in a bit. But for now, if $x_1, x_2, \ldots, x_m = 0$, then:

$$a_{11} 0 + a_{12} 0 + \cdots a_{1m} 0 - b_1 = -t_1$$
$$a_{21} 0 + a_{22} 0 + \cdots a_{2m} 0 - b_2 = -t_2$$
$$\vdots$$
$$a_{n1} 0 + a_{n2} 0 + \cdots a_{nm} 0 - b_n = -t_n$$

$$c_1 0 + c_2 0 + \cdots c_m 0 - d = f.$$

Activity 2.1.5 Consider the region in $\mathbb{R}^2$ bound by $3x + 5y \leq 8, 2x + y \leq 3$ and $x, y \geq 0$. Let $3x + 5y - 8 = -t_1$ and $2x + y - 3 = -t_2$.

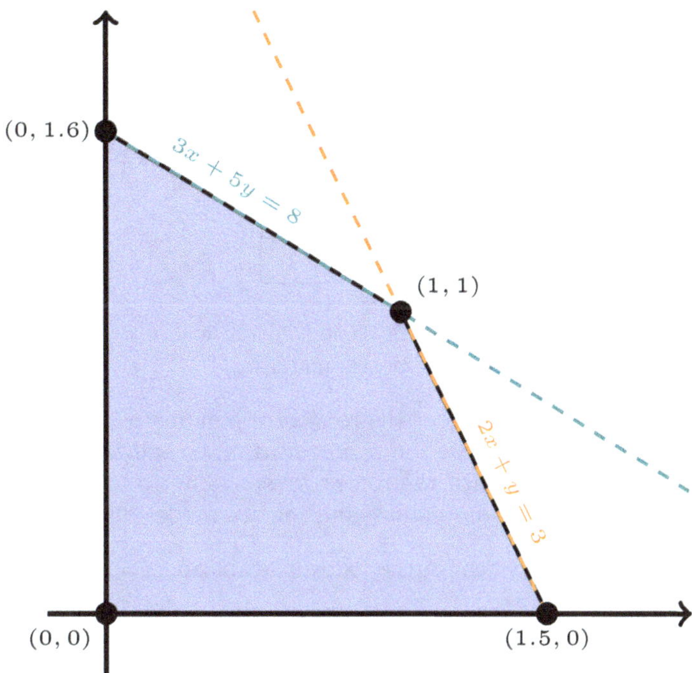

(a) If $t_1 = 0$, what point(s) in $\mathbb{R}^2$ satisfy $3x + 5y - 8 = -t_1$?

(b) If $t_2 = 0$, what point(s) in $\mathbb{R}^2$ satisfy $2x + y - 3 = -t_2$?

(c) What point(s) in $\mathbb{R}^2$ satisfy both $t_1 = 0, t_2 = 0$?

(d) What point(s) in $\mathbb{R}^2$ satisfy both $t_1 = 8, t_2 = 3$?

(e) What point(s) in $\mathbb{R}^2$ satisfy both $t_1 = 0, x = 0$? What about $t_2 = 0, y = 0$?

(f) Consider the interactive below. Drag around the point (x, y). For each variable x, y, t_1, t_2, when are they positive? Negative? Equal to zero?

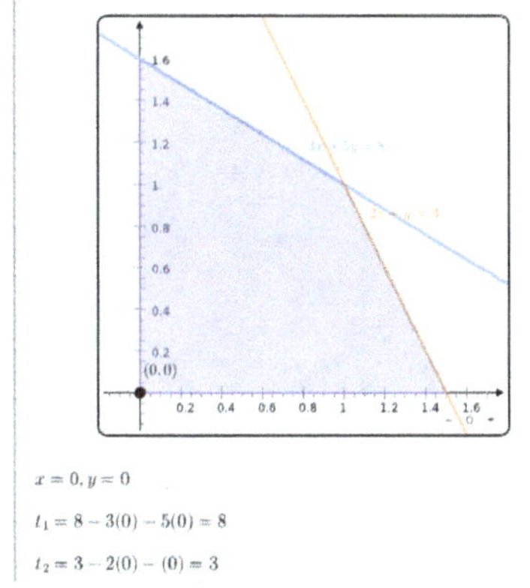

Standalone
Embed

Activity 2.1.6 Suppose a company produces two types of widgets. Widget 1 sells for \$200 and Widget 2 sells for \$150. Each widget takes ingredients A, B and C. Widget 1 needs 1 unit of A, 2 units of B and 2 units of C. Widget 2 needs 2 units of A, 2 units of B and 1 unit of C. The company has 20 units of ingredient A, 30 units of B and 25 units of C.

(a) Set up the canonical maximization problem for the information given above and record it in the following tableau:

x_1	x_2	-1	
?	?	?	$= -t_A$
?	?	?	$= -t_B$
?	?	?	$= -t_C$
?	?	?	$= f$

(b) Recall that this tableau has a basic solution where x_1 and $x_2 = 0$. If we slightly increase x_1 and x_2, do we increase f?

(c) Let's increase x_1. Consider the row corresponding to t_C. Take the associated equality and solve for x_1 in terms of x_2 and t_C. Then, for each equality associated with the rows corresponding to t_A, t_B, f, replace x_1 with the value you found above and rewrite the left hand sides.

(d) Record this new system in the following tableau:

t_C	x_2	-1	
?	?	?	$= -t_A$
?	?	?	$= -t_B$
?	?	?	$= -x_1$
?	?	?	$= f$

(e) Recall that this new tableau has a basic solution where t_C and $x_2 = 0$. What is x_1? Where in $\mathbb{R}^2$ is this solution?

(f) If we increase t_C from 0, do we increase f? What about increasing x_2?

(g) Let's increase x_2. Take the row corresponding to t_B and repeat Tasks **(c)** and **(d)** to obtain a tableau of the form:

t_C	t_B	-1	
?	?	?	$= -t_A$
?	?	?	$= -x_2$
?	?	?	$= -x_1$
?	?	?	$= f$

(h) What point in $\mathbb{R}^2$ represents the basic solution for this new tableau?

(i) If we increase t_C from 0, do we increase f? What about t_B?

(j) Consider the plot of the feasible region for this problem. What exactly, geometrically, did we end up doing in this activity?

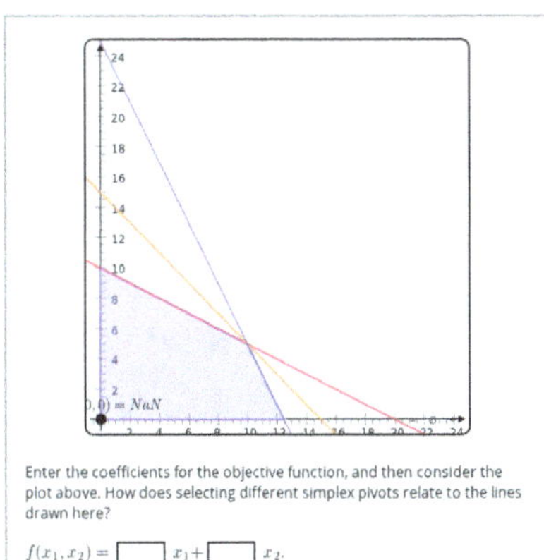

Enter the coefficients for the objective function, and then consider the plot above. How does selecting different simplex pivots relate to the lines drawn here?

$f(x_1, x_2) = \boxed{}\ x_1 + \boxed{}\ x_2.$

Activity 2.1.7 Based on Activity 2.1.6, what would be a reasonable sufficient condition for a feasible tableau to record a basic optimal solution?

A. Some of the $c_j \leq 0$.

B. All of the $c_j \leq 0$.

C. Some of the $c_j \geq 0$.

D. All of the $c_j \geq 0$.

E. Some of the $b_i \leq 0$.

F. All of the $b_i \leq 0$.

G. Some of the $b_i \geq 0$.

H. All of the $b_i \geq 0$.

Activity 2.1.8 Consider the tableau:

x_1	x_2	-1	
a_{11}	a_{12}	b_1	$= -t_1$
a_{21}	a_{22}	b_2	$= -t_2$
c_1	c_2	d	$= f$

along with the corresponding system of equations.

(a) Solve for x_1 in terms of t_1 and x_2.

(b) In each of the other two equalities, replace x_1 with the expression we found above and simplify.

(c) Record this new system in the following tableau:

t_1	x_2	-1	
?	?	?	$= -x_1$
?	?	?	$= -t_2$
?	?	?	$= f$

Definition 2.1.9 The following is the procedure for a **pivot transformation**:

1. Pick a nonzero entry p in the tableau but not in the objective row or constraint column.

2. Transpose the decision and slack variables corresponding to the position of p.

3. Replace p by $1/p$.

4. Replace each entry s in the same row as p (but not p) with s/p.

5. Replace each entry r in the same column as p (but not p) with $-r/p$.

6. Each entry q not in the same row or column as p but in the same column as s (which is in the same row as p) and in the same row as r (which is in the same column as p) is replaced with $\frac{pq-rs}{p}$.

$\Diamond$

Activity 2.1.10 Consider the problem: maximize $f(x,y) = 5x_1 - x_2$ subject to:

$$-7x_1 + x_2 \le 0$$
$$x_1 + 2x_2 \le 30$$
$$x_1 - 2x_2 \le 0$$
$$3x_1 - x_2 \le 20$$
$$x_1, x_2 \ge 0$$

which we may encode as:

x_1	x_2	-1	
-7	1	0	$= -t_1$
1	2	30	$= -t_2$
1	-2	0	$= -t_3$
3	-1	20	$= -t_4$
5	-1	0	$= f$

(a) Pivot on the entry with $*$ (Keep track of the decision and slack variables.)

x_1	x_2	-1	
-7	1	0	$= -t_1$
1	2	30	$= -t_2$
1^*	-2	0	$= -t_3$
3	-1	20	$= -t_4$
5	-1	0	$= f$

(b) Pivot on the entry with $*$ (Keep track of the decision and slack variables.)

$?$	$?$	-1	
$?$	$?$	$?$	$= -?$
$?$	$?$	$?$	$= -?$
$?$	$?$	$?$	$= -?$
$?$	$?^*$	$?$	$= -?$
$?$	$?$	$?$	$= f$

(c) Pivot on the entry with $*$ (Keep track of the decision and slack variables.)

$$
\begin{array}{ccc}
? & ? & -1
\end{array}
$$

?	?	?	$= -?$
?*	?	?	$= -?$
?	?	?	$= -?$
?	?	?	$= -?$
?	?	?	$= f$

(d) Look at your decision variables. Which two lines are we currently sitting on?

(e) How do we know the basic solution we now have is an optimal solution?

(f) Confirm your solution geometrically:

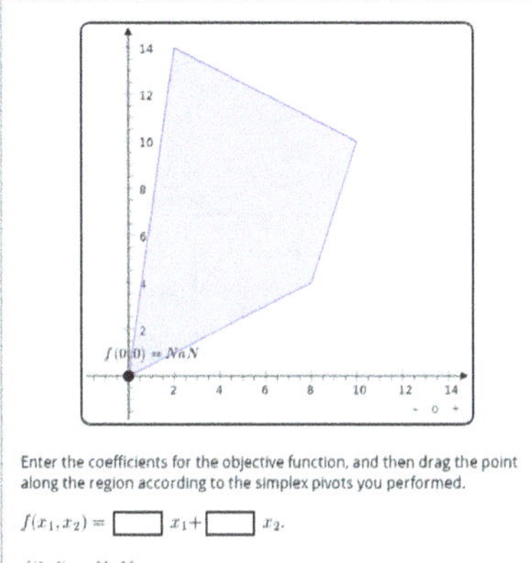

Remark 2.1.11 One may observe, this pivot operation is not difficult, but rather tedious, and it is easy to make a minor error that derails this process. Doing multiple pivots in a row compounds this issue. It is like a more tedious version of row reduction and Gaussian elimination. Thus an interactive pivoter has been provided in Appendix B. Hopefully this improves everyone's lives.

Activity 2.1.12 Suppose a company produces two types of widgets. Widget 1 sells for \$200 and Widget 2 sells for \$150. Each widget takes ingredients A, B and C. Widget 1 needs 1 unit of A, 2 units of B and 2 units of C. Widget 2 needs 2 units of A, 2 units of B and 1 unit of C. The company has 20 units of ingredient A, 30 units of B and 25 units of C.

Now, the company wants to assign values y_A, y_B, y_C to the three ingredients. The values for each should be enough so that in a disaster, the potential revenue is recovered, i.e.:

$$y_A + 2y_B + 2y_C \geq 200$$
$$2y_A + 2y_B + y_C \geq 150$$

Of course, the values shouldn't be negative, so $y_A, y_B, y_C \geq 0$. But, the higher we value the ingredients, the greater the insurance premiums will be, so we need to minimize $g(y_A, y_B, y_C) = 20y_A + 30y_B + 25y_C$.

We can convert this into a max problem to solve, but we can also record it in the following tableau:

y_A	1	2	20
y_B	2	2	30
y_C	2	1	25
-1	200	150	0
	$= s_1$	$= s_2$	$= g$

(a) For each column j, consider the equality recorded by this tableau: $a_{1j}y_A + a_{2j}y_B + a_{3j}y_C + (-1)c_j = s_j$. What must be true about s_j in order for $a_{1j}y_A + a_{2j}y_B + a_{3j}y_C \geq c_j$?

A. $s_j \geq 0$.

B. $s_j \leq 0$.

C. $s_j = 0$.

D. Nothing may be determined about s_j.

(b) Pivot on the entry with the $*$:

y_A	1	2	20
y_B	2	2	30
y_C	2*	1	25
-1	200	150	0
	$= s_1$	$= s_2$	$= g$

(c) Pivot on the entry with the $*$:

y_A	?	?	?
y_B	?	?*	?
s_1	?	?	?
-1	?	?	0
	$= y_C$	$= s_2$	$= g$

(d) Compare this basic solution and tableau to the final solution in Activity 2.1.6. What do you notice?

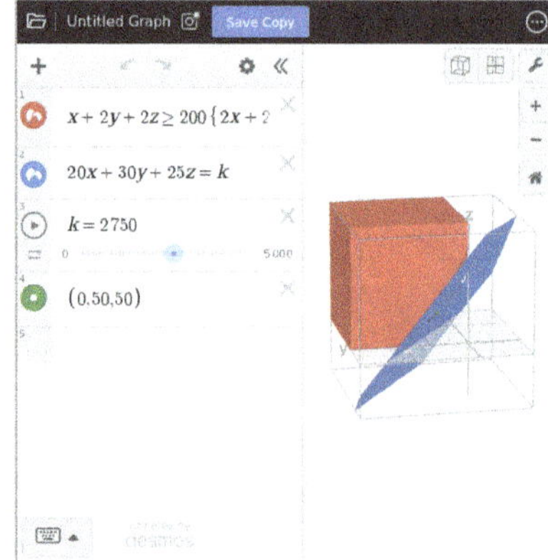

Standalone

2.2 The Simplex Algorithm for Canonical Maximization

Having established the pivot in Section 2.1, we use geometry to determine where to pivot to find optimal solutions. We also discuss how to address potential obstacles.

2.2.1 Basic Feasible Maximization

Activity 2.2.1

(a) Consider the following tableau:

x_1	t_2	t_1	x_4	-1	
2	4	6	5	3	$= -x_3$
-3	0	$\frac{2}{3}$	-2	-1	$= -x_2$
$\frac{17}{4}$	-1	4	$\frac{2}{3}$	-2	$= -t_3$
5	-1	0	3	-244	$= f$

Note that $x_1, t_2, t_1, x_4 = 0$ for the basic solution of this tableau. Write out the corresponding system of canonical inequalities. Are they all satisfied?

(b) Which of the following is a necessary and sufficient condition for a Tucker tableau to have a feasible basic solution?

   A. Some of the $c_j \leq 0$.   E. Some of the $b_i \leq 0$.

   B. All of the $c_j \leq 0$.   F. All of the $b_i \leq 0$.

   C. Some of the $c_j \geq 0$.   G. Some of the $b_i \geq 0$.

   D. All of the $c_j \geq 0$.   H. All of the $b_i \geq 0$.

Activity 2.2.2 Consider the following tableau:

x_3	t_3	x_2	t_1	-1	
-4	25	-3	2	18	$= -t_2$
-1	2	2	-1	22	$= -x_1$
-2	0	-1	-1	10	$= -x_4$
2	0	-1	1	-156	$= f$

Note that this tableau records a basic solution where $x_3, t_3, x_2, t_1 = 0$.

(a) Why is the basic solution feasible?

(b) Which of the following x_3, t_3, t_1, x_2 could we increase from zero to increase our objective function value?

(c) If we increase t_1 to 5, are all our inequalities satisfied? What about 10? 100?

(d) If we increase x_3 to 5, are all our inequalities satisfied? What about 10? 100?

(e) What is the largest value we could increase t_1 while satisfying our 3 inequalities? What about x_3?

(f) Find a sufficient condition for a feasible tableau for a canonical maximization problem to have an unbounded feasible region.

A. All of the $a_{ij} \leq 0$.

B. There is a row i so that all of the $a_{ij} \leq 0$ in that row.

C. There is a column j so that all of the $a_{ij} \leq 0$ in that column.

D. All of the $c_j \leq 0$.

E. All of the $b_i \leq 0$.

Activity 2.2.3 So we have found a reasonable condition for a canonical maximization problem to have unbounded feasible region. We now consider objective functions.

(a) Before reading ahead, discuss whether or not the statements "a canonical maximization problem has unbounded feasible region" and "a canonical maximization problem has unbounded objective function" are logically equivalent. If they are, why? If not, does one imply the other, or can each statement hold while the other statement is false?

(b) Consider the following tableau:

x_1	x_2	-1	
-2	1	12	$= -t_1$
-3	4	10	$= -t_2$
-4	1	0	$= f$

Our previous discussion leads us to believe that x_1 may be increased as much as we please, and our two inequalities will remain satisfied. Yet consider a sketch of the region and function:

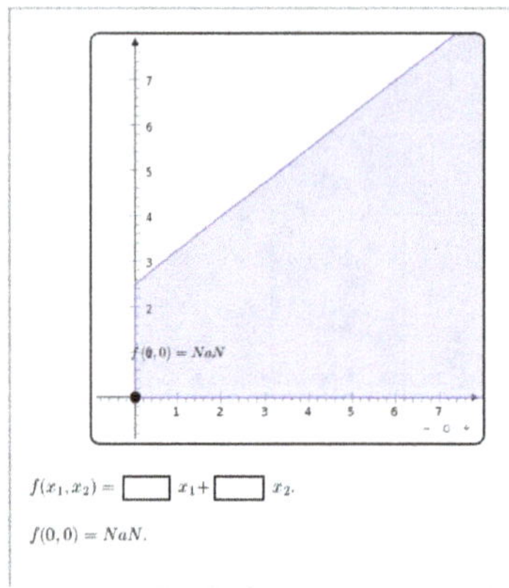

It turns out this system still achieves a maximum solution. Why does this not contradict our previous work?

(c) Which of the following is a reasonable sufficient condition for a feasible tableau for a canonical maximization problem to have an unbounded objective function.

A. All of the $a_{ij} \leq 0$.

B. There is a row i so that all of the $a_{ij} \leq 0$ in that row and $b_i > 0$.

C. There is a row i so that all of the $a_{ij} \leq 0$ in that row and $b_i < 0$.

D. There is a column j so that all of the $a_{ij} \leq 0$ in that column and $c_j > 0$.

E. There is a column j so that all of the $a_{ij} \leq 0$ in that column and $c_j < 0$.

Activity 2.2.4 Consider the canonical maximization problem and basic solution encoded by the following tableau:

x_1	x_2	-1	
3	3	13	$= -t_1$
2	3	12	$= -t_2$
9	6	32	$= -t_3$
7	8	32	$= f$

(a) Let's say we wanted to increase x_2 from 0. What is the largest we could increase x_2 to while satisfying all 3 inequalities?

(b) Consider a sketch of our feasible region:

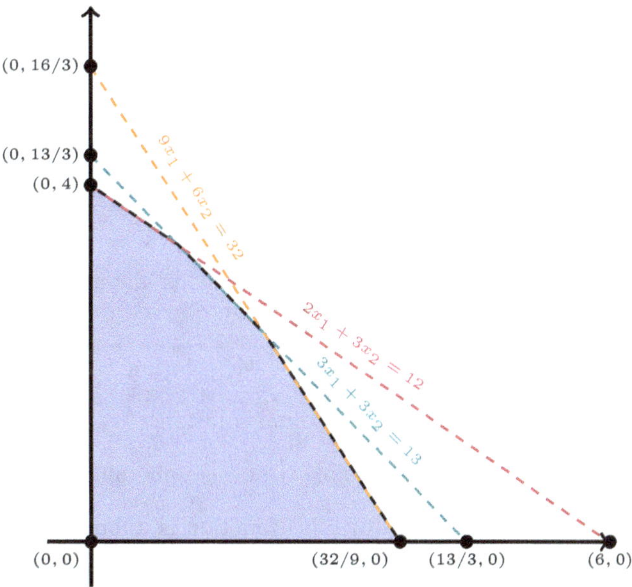

We could choose to pivot x_2 with either t_1, t_2, t_3. Only one of these would result in a feasible solution. Looking graphically, which is it?

(c) What's the connection between the two above tasks?

(d) For a canonical linear problem and basic solution encoded by a Tucker tableau:

$$
\begin{array}{|cccc|c|}
\hline
x_1 & x_2 & \cdots & x_m & -1 \\
\hline
a_{11} & a_{12} & \cdots & a_{1m} & b_1 & = -t_1 \\
a_{21} & a_{22} & \cdots & a_{2m} & b_2 & = -t_2 \\
\vdots & \vdots & \ddots & \vdots & \vdots & \vdots \\
a_{n1} & a_{n2} & \cdots & a_{nm} & b_n & = -t_n \\
\hline
c_1 & c_2 & \cdots & c_m & d \\
\hline
\end{array}
$$

If we pivot in column j, which row i should we choose?

A. Any row i.

B. Any i as long as $a_{ij} > 0$.

C. The row i where $\frac{b_i}{a_{ij}}$ is minimized.

D. The row i where $\frac{c_j}{a_{ij}}$ is minimized.

E. The row i where $\frac{b_i}{a_{ij}}$ is minimized of of the rows where $a_{ij} > 0$.

F. The row i where $\frac{c_j}{a_{ij}}$ is minimized of of the rows where $a_{ij} > 0$.

(e) Based on the observations above, what should be the pivot entry for the following tableau?

$$
\begin{array}{|ccc|}
\hline
x_1 & x_2 & -1 \\
\hline
3 & 4 & 10 & = -t_1 \\
-4 & 3 & 8 & = -t_2 \\
2 & 1 & 5 & = -t_3 \\
\hline
3 & -2 & 0 & = f \\
\hline
\end{array}
$$

Activity 2.2.5 For each of the following tableaus, determine which, if any, of the following are true:

- The tableau records an infeasible basic solution.

- The tableau records a basic optimal solution.

- The tableau tells us the feasible region is unbounded.

- The tableau tells us the objective function is unbounded.

If a tableau is feasible but not optimal, determine the legit pivot points.

$$
\text{(a)} \quad
\begin{array}{|cccc|}
\hline
? & ? & ? & -1 \\
\hline
1 & 1 & 4 & 4 & = -? \\
2 & 2 & 3 & -1 & = -? \\
1 & 4 & 1 & 3 & = -? \\
\hline
1 & -2 & 2 & -5 & = f \\
\hline
\end{array}
$$

(b)

?	?	?	−1	
−3	1	0	5	= −?
5	−3	8	4	= −?
2	4	2	2	= −?
−2	3	0	2	= f

(c)

?	?	?	−1	
3	4	−1	2	= −?
6	−3	2	1	= −?
4	−2	0	3	= −?
−4	−2	0	17	= f

(d)

?	?	?	−1	
−4	3	−2	2	= −?
0	−1	4	0	= −?
−1	4	0	5	= −?
−2	3	2	−34	= f

(e)

?	?	?	−1	
1	−1	0	−2	= −?
−1	−3	8	4	= −?
3	1	2	2	= −?
−2	−3	−5	9	= f

(f)

?	?	?	−1	
3	−2	6	3	= −?
4	−3	−3	0	= −?
−1	2	5	1	= −?
−1	−4	−2	−54	= f

(g)

?	?	?	−1	
0	1	0	4	= −?
2	2	8	6	= −?
2	0	3	3	= −?
5	3	−4	−23	= f

With all this, we finally may define our *Simplex Algorithm for Maximum Basic Feasible Tableaus*.

Definition 2.2.6 The Simplex Algorithm for Maximum Basic Feasible Tableaus

1. Ensure that each $b_i \geq 0$, that is that the solution is feasible.

2. IF every $c_j \leq 0$: STOP since the current solution is optimal.

3. PICK a column j such that $c_j > 0$.

4. IF each $a_{1j}, a_{2j}, \ldots, a_{mj} \leq 0$: STOP since the objective function is unbounded.

5. PICK a row i that minimizes $\{b_i/a_{ij} : a_{ij} > 0\}$.

6. PIVOT on a_{ij}.

7. GOTO 1.

$\Diamond$

Activity 2.2.7 Recall the canonical tableau:

x_1	x_2	$\cdots$	x_m	-1	
a_{11}	a_{12}	$\cdots$	a_{1m}	b_1	$= -t_1$
a_{21}	a_{22}	$\cdots$	a_{2m}	b_2	$= -t_2$
$\vdots$	$\vdots$	$\ddots$	$\vdots$	$\vdots$	$\vdots$
a_{n1}	a_{n2}	$\cdots$	a_{nm}	b_n	$= -t_n$
c_1	c_2	$\cdots$	c_m	d	$= f$

Suppose that the basic solution encoded by the tableau were feasible, and that after pivoting on a_{ij} according to Definition 2.2.6 we had that $d \to d'$. Show that $d' \leq d$.

Activity 2.2.8 Apply Definition 2.2.6 to the following tableaus.

(a)

x_1	x_2	-1	
2	1	8	$= -t_1$
1	2	10	$= -t_2$
30	50	0	$= f$

(b)

x_1	x_2	-1	
-1	1	1	$= -t_1$
1	-1	3	$= -t_2$
1	2	0	$= f$

(c)

x_1	x_2	x_3	x_4	-1	
1	1	1	0	1	$= -t_1$
0	1	1	0	4	$= -t_2$
-1	0	-1	1	1	$= -t_3$
-1	-1	1	1	0	

2.2.2 Basic Infeasible Maximization

Activity 2.2.9 Each of the following tableaus records the origin as a basic solution, which is not feasible. Determine geometrically which of these tableau record a problem which has an optimal solution. Then answer the follow-up questions.

(a)

	x_1	x_2	-1	
	-2	-1	-1	$=-t_1$
	1	5	15	$=-t_2$
	2	3	18	$=-t_3$
	3	5	0	$=f$

(b)

	x_1	x_2	-1	
	-2	3	1	$=-t_1$
	2	-6	-5	$=-t_2$
	5	1	7	$=-t_3$
	5	1	0	$=f$

(c) For the one problem where the origin is not feasible, but the problem has an optimal solution, what pivot would result in a feasible solution?

(d) If $b_{i*} < 0$, what should be true about a_{ij*} so that b_{i*} is positive afterwards?

(e) Consider the inequality $2x_1 - 3x_2 - 4x_3 \le -5$. Is there a solution which satisfies this where each $x_i \ge 0$?

(f) Consider the inequality $5x_1 + x_2 + 2x_3 \le -10$. Is there a solution which satisfies this where each $x_i \ge 0$?

Activity 2.2.10 We now want to concoct a scheme to turn all the b_i nonnegative, if possible.

Consider the following tableau:

	$?$	$?$	$?$	-1	
	a_{11}	a_{12}	a_{13}	b_1	$=-?$
	a_{21}	a_{22}	a_{23}	b_2	$=-?$
	a_{31}	a_{32}	a_{33}	b_3	$=-?$
	a_{41}	a_{42}	a_{43}	b_4	$=-?$
	c_1	c_2	c_3	0	$=f$

Suppose that $b_2 < 0, b_3, b_4 > 0$. We want to turn b_2 positive while making sure b_3, b_4 stay positive.

(a) Suppose each $a_{2j} \ge 0$, what can we say about the associated problem?

(b) Let's then say $a_{22} < 0$. If we pivot on a_{22} do b_2, b_3, b_4 increase or decrease? Are any now guaranteed to be positive or negative?

(c) Let's then say $a_{32} > 0$. If we pivot on a_{32} do b_2, b_3, b_4 increase or decrease? Are any now guaranteed to be positive or negative?

(d) Let's then say $a_{42} < 0$. If we pivot on a_{42} do b_2, b_3, b_4 increase or decrease? Are any now guaranteed to be positive or negative?

(e) Who should we pivot on?

    A. Any a_{2k}, where $k \ge 2$.

  B. All of the $a_{2k}, k \geq 2$ such that b_k/a_{2k} is minimized.

  C. All of the $a_{2k}, k \geq 2$ such that if $k > 2$ then $a_{2k} > 0$ and b_k/a_{2k} is minimized.

(f) Note that we have no idea if b_1 is positive or negative before we started, much less after. Why don't we care?

With this, we may define our algorithm for potential nonfeasible tableau.

Definition 2.2.11 The Simplex Algorithm for Maximum Tableaus

1. IF each $b_i \geq 0$, GOTO 8.

2. PICK the largest i such that $b_i < 0$.

3. IF $a_{i1}, a_{i2}, \ldots, a_{in} \geq 0$: STOP since the problem in infeasible.

4. PICK a column j such that $a_{ij} < 0$.

5. PICK a row $p \geq i$ so that $\min_{k>i} \left(\{b_i/a_{ij}\} \cup \{b_k/a_{kj} : a_{kj} > 0\} \right) = b_p/a_{pj}$.

6. PIVOT on a_{pj}.

7. GOTO 1.

8. APPLY Definition 2.2.6.

$\Diamond$

Activity 2.2.12 Apply Definition 2.2.11 to the following tableau:

(a)

	?	?	−1	
	−1	−2	−3	$= -?$
	1	1	2	$= -?$
	2	−4	0	$= f$

(b)

	?	?	−1	
	−3	−2	−1	$= -?$
	3	5	20	$= -?$
	2	0	4	$= -?$
	3	4	0	$= f$

2.2.3 The Simplex Algorithm for Canonical Minimization

Definition 2.2.13 Consider a canonical minimization problem:

$$\textbf{Minimize: } g(\mathbf{x}) = c_1 x_1 + c_2 x_2 + \cdots c_n x_n - d = \left(\sum_{j=1}^{n} c_j x_j \right) - d$$

$$\textbf{subject to: } a_{11}x_1 + a_{12}x_2 + \cdots a_{1n}x_n \geq b_1$$

$$a_{21}x_1 + a_{22}x_2 + \cdots a_{2n}x_n \geq b_2$$

$$\vdots$$

$$a_{m1}x_1 + a_{m2}x_2 + \cdots a_{mn}x_n \geq b_m$$

$$x_1, x_2, \ldots, x_n \geq 0$$

where $a_{ij}, b_i, c_j, d \in \mathbb{R}$.

This problem may be recorded in the following tableau:

x_1	a_{11}	a_{12}	$\cdots$	a_{1m}	c_1
x_2	a_{21}	a_{22}	$\cdots$	a_{2m}	c_2
$\vdots$	$\vdots$	$\vdots$	$\ddots$	$\vdots$	$\vdots$
x_n	a_{n1}	a_{n2}	$\cdots$	a_{nm}	c_n
-1	b_1	b_2	$\cdots$	b_m	d
	$= t_1$	$= t_2$	$\cdots$	$= t_m$	$= g$

The **negative transpose** of this tableau is:

x_1	x_2	$\cdots$	x_n	-1	
$-a_{11}$	$-a_{21}$	$\cdots$	$-a_{n1}$	$-b_1$	$= -t_1$
$-a_{12}$	$-a_{22}$	$\cdots$	$-a_{n2}$	$-b_2$	$= -t_2$
$\vdots$	$\vdots$	$\ddots$	$\vdots$	$\vdots$	$\vdots$
$-a_{1m}$	$-a_{2m}$	$\cdots$	$-a_{nm}$	$-b_m$	$= -t_m$
$-c_1$	$-c_2$	$\cdots$	$-c_n$	$-d$	

$\Diamond$

Activity 2.2.14

(a) Show that this negative transpose of a canonical minimization tableau encodes a problem that has the same feasible region and optimal solution as the original minimization problem.

(b) Describe a procedure to utilize Definition 2.2.11 to solve a minimization problem.

(c) Use this procedure to solve the problem presented in Activity 1.2.2.

2.3 Cycling

In this section, we discuss a potentially serious hurdle to using the Simplex Algorithm, and show how to avoid it completely.

Activity 2.3.1 Consider the following canonical maximization tableau:

x_1	x_2	x_3	x_4	-1	
-12.5	-2	12.5	1	0	$= -t_1$
1	0.24	-2	-0.24	0	$= -t_2$
4	1.92	-16	-0.96	1	$= f$

(a) Perform the following sequence of pivots, ensure each time that it is a valid pivot according to Definition 2.2.6, and keep track of the variables.

?	?	?	?	-1	
-12.5	-2	12.5	1	0	$= -?$
1^*	0.24	-2	-0.24	0	$= -?$
4	1.92	-16	-0.96	1	$= f$

?	?	?	?	-1	
?	?*	?	?	?	$= -?$
?	?	?	?	?	$= -?$
?	?	?	?	?	$= f$

?	?	?	?	-1	
?	?	?	?	?	$= -?$
?	?	?*	?	?	$= -?$
?	?	?	?	?	$= f$

?	?	?	?	-1	
?	?	?	?*	?	$= -?$
?	?	?	?	?	$= -?$
?	?	?	?	?	$= f$

?	?	?	?	-1	
?	?	?	?	?	$= -?$
?*	?	?	?	?	$= -?$
?	?	?	?	?	$= f$

?	?	?	?	-1	
?	?*	?	?	?	$= -?$
?	?	?	?	?	$= -?$
?	?	?	?	?	$= f$

(b) Compare your final tableau to the initial tableau. Are there any similarities?

Activity 2.3.2 Consider the canonical linear optimization problem: Maximize $f(x, y) = x + y$, subject to

$$-2x + y \leq 0$$
$$x - 2y \leq 0$$
$$x + y \leq 4$$
$$x, y \geq 0.$$

Consider a sequence of pivots swapping $x \leftrightarrow t_2$, $y \leftrightarrow t_1$ $t_1 \leftrightarrow y$, $t_2 \leftrightarrow x$. In each of these cases, what is the basic solution recorded by the tableau?

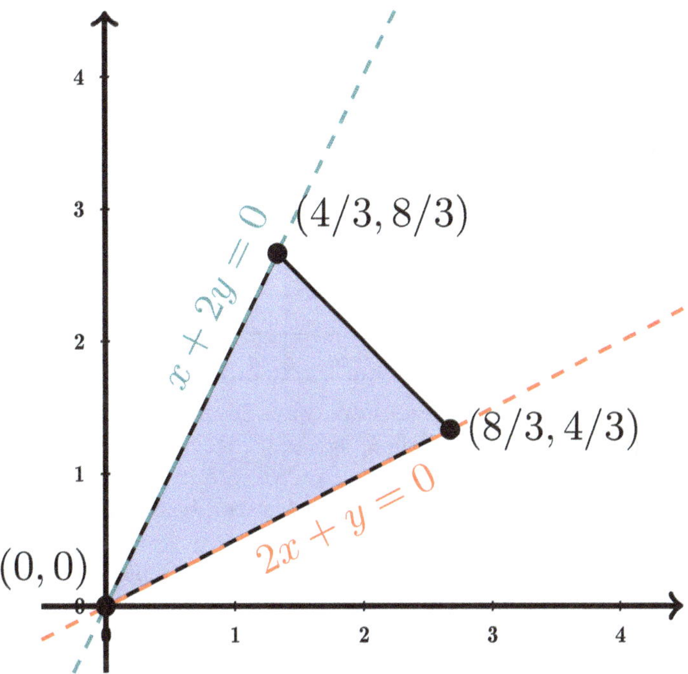

Definition 2.3.3 If a series of pivots in accordance to Definition 2.2.6 allows one to return to a set of basic variables achieved earlier in the algorithm, we call this phenomenon **cycling**. ◇

Definition 2.3.4 If a pivot on a tableau gives a new tableau corresponding to the same basic solution, we call the pivot a **degenerate** pivot. ◇

Cycling is bad, since potentially this allows the Simplex Algorithm to not terminate. Fortunately, there is a solution to this issue.

Theorem 2.3.5 Bland's Anticycling Rules. *List all variables, independent and dependent in the initial tableau. Then consider the following rules:*

1. *Whenever there is more than one possible choice for a pivot row according to Definition 2.2.6: select the row corresponding to the variable closest to the front of the initial list.*

2. *Whenever there is more than one possible choice for a pivot column according to Definition 2.2.6: select the column corresponding to the variable closest to the front of the initial list.*

Then the Simplex Algorthm will not cycle.

Before we proceed with the proof, we require a technical definition.

Definition 2.3.6 We call the **dictionary** for a basis the system of equations corresponding to that basis. So the dictionary for a basis B_i is D_i which we can write with constraint equalities:

$$x_k = b_k - \sum_{x_j \notin B_i} a_{kj} x_j \text{ for } k \in B_i$$

and objective function equality:

$$f = \sum_{x_j \notin B_i} c_j x_j - d.$$

$\Diamond$

Proof. For the sake of notation, we denote the initial tableau:

$$
\begin{array}{cccc|c|l}
x_1 & x_2 & \cdots & x_m & -1 & \\
\hline
a_{11} & a_{12} & \cdots & a_{1m} & b_1 & = -x_{m+1} \\
a_{21} & a_{22} & \cdots & a_{2m} & b_2 & = -x_{m+2} \\
\vdots & \vdots & \ddots & \vdots & \vdots & \vdots \\
a_{n1} & a_{n2} & \cdots & a_{nm} & b_n & = -x_{m+n} \\
\hline
c_1 & c_2 & \cdots & c_m & d & = f
\end{array}
$$

and order the variables in the associated way. Suppose we follow Bland's rules and still have cycling. That is, there is a sequence of pivots and bases (set of basic variables) $B_1 \to B_2 \to \cdots B_\ell \to B_1$. We call a variable x_j **fickle** if x_j is in one, but not all of the bases. That is, it leaves a basis, and re-enters it later during the cycle.

Note that in order for cycling to occur, each pivot must be degenerate (ask yourself why?). So if x_j is fickle, $x_j = 0$ for each basic solution during the cycle.

Since the number of variables is finite, the set of fickle variables is also finite, and let x_t denote the fickle variable with largest subscript. Suppose that $x_t \in B_i, x_t \notin B_{i+1}$ (why must such a B_i exist?). Denote the entering variable from $B_i \to B_{i+1}$ with x_s.

$$
\begin{array}{cccc|c|l}
\cdots & x_s & \cdots & -1 & \\
\hline
\cdots & \cdots & \cdots & \vdots & \\
\vdots & a_{ts} & \vdots & 0 & = -x_t \\
\cdots & \cdots & \cdots & \vdots & \\
\hline
\cdots & c_s & \cdots & d & = f
\end{array}
\qquad
\begin{array}{cccc|c|l}
\cdots & x_t & \cdots & -1 & \\
\hline
\cdots & \cdots & \cdots & \vdots & \\
\vdots & ? & \vdots & 0 & = -x_s \\
\cdots & \cdots & \cdots & \vdots & \\
\hline
\cdots & ? & \cdots & d & = f
\end{array}
$$

Note that x_s must also be fickle (why?) and so $s < t$. We also see that since the above pivot was valid, we must have that $c_s > 0, a_{ts} > 0$ and since the pivot was degenerate, we have $b_t = 0$.

Now, because we have cycling, we must have that x_t re-enters the basis at some point:

$$
\begin{array}{|ccc|c|}
\hline
\cdots & x_t & \cdots & -1 \\
\hline
\cdots & \cdots & \cdots & \vdots \\
\vdots & ? & \vdots & 0 \\
\cdots & \cdots & \cdots & \vdots \\
\hline
\cdots & c_t^* & \cdots & d \\
\hline
\end{array}
\begin{array}{l}
{} \\
{} \\
= -x_? \\
{} \\
= f
\end{array}
\qquad
\begin{array}{|ccc|c|}
\hline
\cdots & x_? & \cdots & -1 \\
\hline
\cdots & \cdots & \cdots & \vdots \\
\vdots & ? & \vdots & 0 \\
\cdots & \cdots & \cdots & \vdots \\
\hline
\cdots & ? & \cdots & d \\
\hline
\end{array}
\begin{array}{l}
{} \\
{} \\
= -x_t \\
{} \\
= f
\end{array}
$$

Note that for this pivot to be valid, we must have that $c_t^* > 0$. If we let D_ℓ denote the dictionary before x_t re-enters the basis, we have:

$$f = \sum_{x_j \notin B_\ell} c_j^* x_j - d.$$

So note that the solution space to the system of equations for both these dictionaries are the same. So we have a solution for D_i (not necessarily feasible ie nonnegativity may fail) that must also be a solution to D_ℓ:

$$
\begin{aligned}
x_s &= Z \\
x_j &= 0 \text{ for } x_j \notin B_i, x_j \neq x_s \\
x_k &= b_k - a_{ks}Z \text{ for } x_k \in B_i \\
f &= c_s Z - d.
\end{aligned}
$$

So we have

$$c_s Z - d = f = c_s^* Z - d + \sum_{x_k \in B_i} c_k^*(b_k - a_{ks}Z)$$

where $c_k^* = 0$ for $x_k \in B_\ell$. So via algebra:

$$\left(c_s - c_s^* + \sum_{x_k \in B_i} c_k^* a_{ks} \right) Z = \sum_{x_k \in B_i} c_k^* b_k.$$

The above equation holds true no matter what Z is. Thus:

$$c_s - c_s^* + \sum_{x_k \in B_i} c_k^* a_{ks} = 0.$$

Note that x_t, NOT x_s is entering the basis. If x_s is already in the basis, $c_s^* = 0$. Otherwise, $c_s^* \leq 0$, otherwise s would have entered by Bland's Anticyling rules. We've also shown $c_s > 0$. Thus

$$\sum_{x_k \in B_i} c_k^* a_{ks} < 0$$

which is only possible if there is some $x_r \in B_i$ such that

$$c_r^* a_{rs} < 0.$$

From this, we know that $c_r^* \neq 0$. So $x_r \notin B_\ell$, but $x_r \in B_i$, so x_r is fickle. Since t is the largest index of a fickle variable, $r < t$. Note that x_r is not entering from B_ℓ, so $c_r^* \leq 0$. Thus $a_{rs} > 0$.

But x_r is fickle, so $b_r = 0$ in D_i. But then, we would have picked x_r, not x_t to leave.

$$
\begin{array}{ccc|c}
\cdots & x_s & \cdots & -1 \\
\hline
\cdots & \cdots & \cdots & \vdots \\
\vdots & a_{ts} & \vdots & 0 \quad = -x_t \\
\vdots & a_{rs} & \vdots & 0 \quad = -x_r \\
\cdots & \cdots & \cdots & \vdots \\
\hline
\cdots & c_s & \cdots & d \quad = f
\end{array}
$$

which is a contradiction. ∎

Activity 2.3.7 If we follow Theorem 2.3.5, then no basis is visited twice during the execution of the Simplex Algorithm. Note that f is nondecreasing with each pivot. Must the Simplex Algorithm terminate? Why?

2.4 Using Sage to Solve Linear Optimization Problems

In practice, most linear algebra computations are done with computers. The presence of tedious technical operations and algorithmic thinking should suggest the same can be done here.

Exploration 2.4.1 Suppose we wanted to solve the following maximization problem by hand:

x_1	x_2	x_3	x_4	x_5	x_6	x_7	x_8	x_9	-1	
9	8	3	2	6	6	2	1	4	38	$= -t_1$
5	3	4	4	7	8	1	3	5	52	$= -t_2$
1	1	2	5	4	5	8	4	4	35	$= -t_3$
1	6	2	2	5	5	4	7	3	76	$= -t_4$
5	7	6	8	9	5	4	3	6	53	$= -t_5$
6	2	6	5	8	3	5	5	6	59	$= -t_6$
3	7	5	4	6	3	5	5	6	50	$= -t_7$
8	3	6	4	3	2	2	6	2	82	$= -t_8$
7	7	2	6	5	3	7	8	7	37	$= -t_9$
7	7	2	6	5	3	7	8	7	0	$= f$

How annoying would this be?

A. Very.

B. Extraordinarily.

C. Horrifically.

D. I have nothing to do for the next hour anyway. Hope I don't forget a minus sign!

Given that we have a developed an algorithm, guaranteed to terminate, using only arithmetic in its steps, it seems reasonable to think this can be done via a computer.

Activity 2.4.2 Let's start simple, suppose we want to solve:

x_1	x_2	-1	
2	-1	5	$= -t_1$
-2	1	2	$= -t_2$
3	2	12	$= -t_3$
5	4	0	$= f$

(a) We can enter the above problem into Sage via:

```
%display typeset
A = ([2, -1], [-2, 1], [3, 2])
b = (5, 2, 12)
c = (5,4)
P = InteractiveLPProblemStandardForm(A, b, c, )
P
```

(b) We can plot the feasible region and objective level curves, since this is a 2d problem:

```
P.plot()
```

(c) We could also encode this problem into a dictionary.

```
D = P.initial_dictionary()
D
```

We will understand that $t_1 = x_3, t_2 = x_4, t_3 = x_5$.

(d) If we want to pivot from x_4 to x_2 we can write that as:

```
D.enter("x2")
D.leave("x4")
D
```

Then we can update the dictionary:

```
D.update()
D
```

We should read this as:

x_1	x_4	-1	
0	1	7	$= -x_3$
-2	1	2	$= -x_2$
7	-2	8	$= -x_5$
13	-4	-8	$= f$

(e) What pivot should we do next? Encode it below by editing the `FIX_ME` values:

```
D.enter("FIX_ME")
D.leave("FIX_ME2")
D.update()
D
```

(f) We can check at any point if we have an optimal solution.

```
D.is_optimal()
```

Activity 2.4.3 So if we want to solve:

x_1	x_2	x_3	x_4	-1	
8	2	4	5	3	$= -t_1$
-4	6	2	7	4	$= -t_2$
2	8	4	3	2	$= -t_3$
1	3	2	1	0	$= f$

We can encode this in:

```
%display typeset
A = ([8, 2, 4, 5], [-4, 6, 2, 7], [2, 8,4,3])
b = (3, 4, 2)
c = (1, 3, 2, 1)
P = InteractiveLPProblemStandardForm(A, b, c,)
P
```

(a) We now encode the above problem in a dictionary.

```
%display typeset
D = P.initial_dictionary()
D
```

(b) We can see who can enter:

```
D.possible_entering()
```

If we say, pick x_2 to enter, see who can legitimately leave:

```
D.enter("x2")
D.possible_leaving()
```

Then select one to leave:

```
D.leave(FIX_ME)
D.update()
D
```

(c) From here, pick another legitimate pivot and perform it:

```
D.enter("FIX_ME")
D.leave("FIX_ME2")
D.update()
D
```

(d) This still seems like a it would be annoying. Why don't we revisit the original problem?

```
%display typeset
P
```

and then just run the Simplex Algorithm:

```
%display typeset
P.run_simplex_method()
```

Activity 2.4.4 Remember that cycling example Activity 2.3.1?

(a) Encode the problem into Sage:

```
%display typeset
A = (FIX_ME)
b = (FIX_ME)
c = (FIX_ME)
P = InteractiveLPProblemStandardForm(A, b, c, )
P
```

(b) Now let's run the Simplex Algorithm to see what the deal is:

```
%display typeset
P.run_simplex_method()
```

Activity 2.4.5 We can now tackle the problem posed in Exploration 2.4.1.

(a) Encode the problem into Sage:

```
%display typeset
A = (FIX_ME)
b = (FIX_ME)
c = (FIX_ME)
P = InteractiveLPProblemStandardForm(A, b, c, )
P
```

(b) We can run the Simplex Algorithm:

```
%display typeset
P.run_simplex_method()
```

(c) We can also just say what the solution is:

```
print(P.optimal_solution())
print(P.optimal_value())
```

2.5 Summary of Chapter 2

We recall that the canonical linear optimization problem

$$\textbf{Maximize: } f(\mathbf{x}) = c_1 x_1 + c_2 x_2 + \cdots c_m x_m - d = \left(\sum_{j=1}^{m} c_j x_j \right) - d$$

$$\textbf{subject to: } a_{1,1} x_1 + a_{1,2} x_2 + \cdots a_{1,m} x_m \leq b_1$$
$$a_{2,1} x_1 + a_{2,2} x_2 + \cdots a_{2,m} x_m \leq b_2$$
$$\vdots$$
$$a_{n,1} x_1 + a_{n,2} x_2 + \cdots a_{n,m} x_m \leq b_n$$
$$x_1, x_2, \ldots, x_m \geq 0.$$

may be written as

$$\textbf{Maximize: } f(\mathbf{x}) = c_1 x_1 + c_2 x_2 + \cdots c_m x_m - d = \left(\sum_{j=1}^{m} c_j x_j \right) - d$$

$$\textbf{subject to: } a_{1,1} x_1 + a_{1,2} x_2 + \cdots a_{1,m} x_m + t_1 = b_1$$
$$a_{2,1} x_1 + a_{2,2} x_2 + \cdots a_{2,m} x_m + t_2 = b_2$$
$$\vdots$$
$$a_{n,1} x_1 + a_{n,2} x_2 + \cdots a_{n,m} x_m + t_n = b_n$$
$$x_1, x_2, \ldots, x_m, t_1, \ldots, t_n \geq 0.$$

where the t_i are called the **slack variables** for each constraint. We refer to the original x_j as **decision variables**. We noted that by their construction, for any point $\mathbf{x} \in \mathbb{R}^m$, t_i is zero if and only if $\mathbf{x}$ lies on bounding hyperplane, and is positive if it is on the "correct" side of the bounding hyperplane.

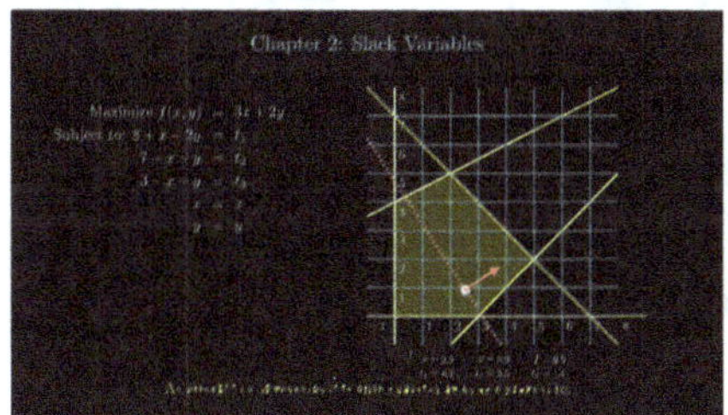

Figure 2.5.1 An introduction to slack variables.

All of this information may be written in a condensed form, the **Tucker tableau.**

x_1	x_2	$\cdots$	x_m	-1	
a_{11}	a_{12}	$\cdots$	a_{1m}	b_1	$= -t_1$
a_{21}	a_{22}	$\cdots$	a_{2m}	b_2	$= -t_2$
$\vdots$	$\vdots$	$\ddots$	$\vdots$	$\vdots$	$\vdots$
a_{n1}	a_{n2}	$\cdots$	a_{nm}	b_n	$= -t_n$
c_1	c_2	$\cdots$	c_m	d	

We also noted that, although our inequalities and objective are written in terms of the x_j, since we have the equality

$$a_{i1} x_1 + \cdots + a_{ij} x_j + \cdots a_{im} x_m = b_i - t_i$$

that we could rewrite as:

$$x_j = \frac{1}{a_{ij}} \left(b_i - t_i - a_{i1} x_i - \cdots - a_{im} x_m \right)$$

provided $a_{ij} \neq 0$. This would allow us to rewrite all the pertinent equalities and inequalities replacing x_j with t_i. This process is called a **pivot transformation** Definition 2.1.9, where now t_i is a decision variable and x_j is a slack variable.

We note that by setting all the decision variables equal to zero, we obtain a potential solution called a **basic solution** and the pivot transformation is really moving from basic solution to basic solution. Assuming that all $b_i \geq 0$, the basic solutions are feasible, and we establish a rule to identify pivots. Picking a $c_j > 0$, we choose a positive entry a_{ij} minimizing the ratio $\frac{b_i}{a_{ij}}$. When each $c_j \leq 0$, we have obtained an optimal solution. This is summarized in the **Simplex Algorithm** Definition 2.2.11.

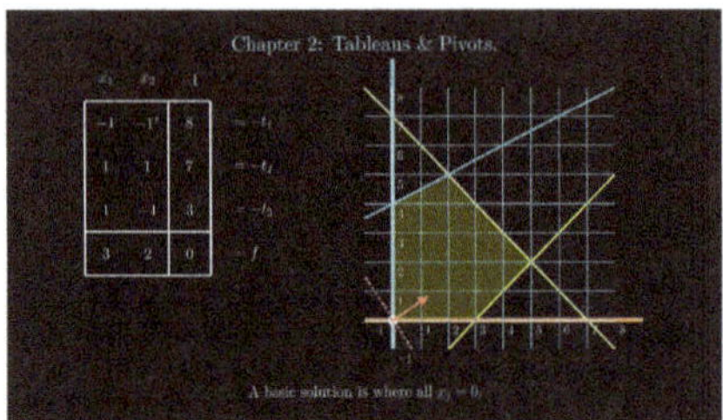

Figure 2.5.2 Pivot transformations and the simplex algorithm.

Note that a feasible region may be unbounded, which we may detect with seeing a column j where each $a_{ij} \leq 0$ but each $b_i \geq 0$. However, this does not mean the objective function is unbounded. This only occurs when $c_j < 0$ as well. Note that this is a sufficient, and not necessary condition.

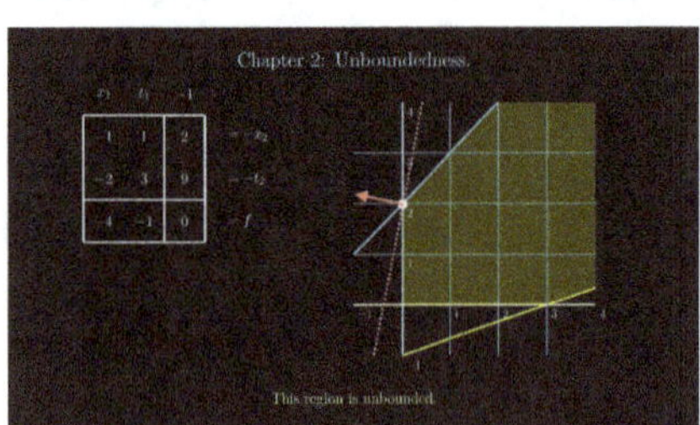

Figure 2.5.3 Unbounded regions and objective functions.

Similarly, whenever $b_i < 0$, the basic solution is not feasible, but the problem may not be infeasible. The problem is only infeasible (i.e. the feasible region is empty) when $a_{ij} \leq 0$ for each a_{ij} in the same row. This is again sufficient but not necessary.

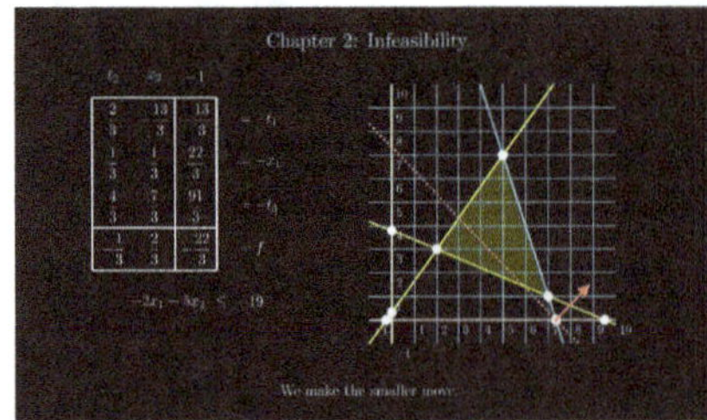

Figure 2.5.4 Infeasible basic solutions and feasible regions.

Finally, it is possible to pivot from basic solution to basic solution, represented as the intersection of different hyperplanes, but without actually changing the point in $\mathbb{R}^m$ in a phenomena called **cycling**. We introduce and prove **Bland's Anticyling Theorem** Theorem 2.3.5 which shows that ordering the variables, and always using the first possible variable to break any ties, we may avoid this issue.

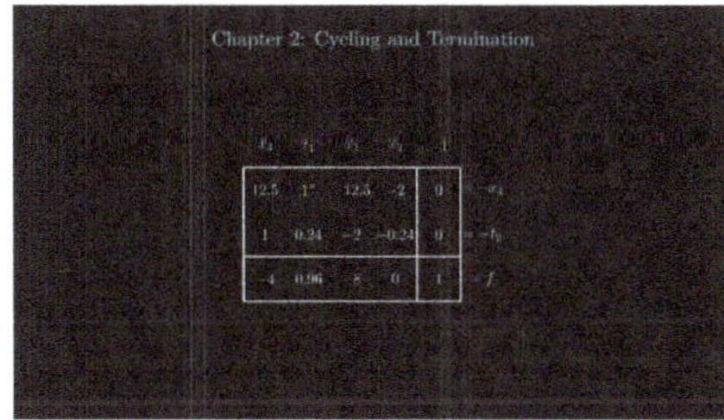

Standalone

Figure 2.5.5 Cycling and Bland's Anticycling Theorem.

2.6 Problems for Chapter 2

Exercises

1. Consider the tableau:

	x_1	x_2	-1	
	3	4	8	$= -t_1$
	1	2	9	$= -t_2$
	5	6	7	$= f$

    (a) Write out the canonical maximization problem encoded by the tableau.

    (b) State the basic solution for this tableau.

    (c) Determine if the basic solution is feasible.

    (d) Pivot on the entry 2.

    (e) Write out the new canonical maximization problem in terms of the nonbasic variables and the new basic solution in terms of x_1, x_2.

2. Consider the tableau:

y_1	3	4	8
y_2	1	2	9
-1	5	6	7
	$= s_1$	$= s_2$	$= g$

    (a) Write out the canonical minimization problem encoded by the tableau.

    (b) State the basic solution for this tableau.

    (c) Determine if the basic solution is feasible.

    (d) Pivot on the entry 3.

    (e) Write out the new canonical minimization problem in terms of the nonbasic variables and the new basic solution in terms of y_1, y_2.

3. For each of the following canonical maximization tableaus:

    (a) Write out the current basic solution.

    (b) Determine if the current basic solution is feasible.

    (c) Determine if the tableau detects that the feasible region is unbounded.

    (d) Determine if the tableau detects that the problem is infeasible. If so, ignore the rest of the prompts.

(e) Determine if the tableau detects that the problem is unbounded. If so, ignore the rest of the prompts.

(f) Determine if the current basic solution is optimal. If so, ignore the rest of the prompts.

(g) Identify all valid pivot entries.

(h) Pivot on the entry corresponding to Bland's Anticycling Theorem Theorem 2.3.5.

(i) Write out the new basic solution.

(a)

x_2	t_1	-1	
3	-2	2	$= -t_2$
-1	4	4	$= -x_1$
-1	2	-12	$= f$

(b)

x_1	t_1	x_2	-1	
0	-2	4	2	$= -x_3$
2	0	-1	4	$= -t_2$
3	-1	0	-12	$= f$

(c)

t_3	x_1	t_2	-1	
-3	5	2	4	$= -x_3$
4	0	1	8	$= -t_1$
6	5	-2	12	$= -x_2$
-1	2	0	10	$= f$

(d)

t_3	x_2	t_1	-1	
$4/5$	-1	3	4	$= -x_1$
-1	$-1/2$	2	$8/5$	$= -t_2$
-2	0	5	$7/2$	$= -x_3$
$5/2$	$-1/2$	3	-120	$= f$

(e)

x_1	x_2	-1	
-1	-2	-3	$= -t_1$
-3	1	5	$= -t_2$
1	1	10	$= -t_3$
3	4	0	$= f$

(f)

t_3	t_1	-1	
0	2	6	$= -t_2$
-1	3	6	$= -x_2$
0	3	10	$= -x_1$
4	-1	-8	$= f$

(g)

t_2	x_1	x_2	-1	
3	2	6	12	$= -x_3$
-1	0	4	8	$= -t_1$
0	2	1	-5	$= -t_3$
-3	0	-1	-18	$= f$

4. For each problem in Exercise 1.5.3 solve these problems using the Simplex Algorithm.

5. Solve the following using the Simplex Algorithm.

(a)

$$\textbf{Maximize: } f(x,y) = -x + 4y$$
$$\textbf{subject to: } x - y \geq 2$$
$$2x - 3y \leq 10$$
$$5x + 4y \leq 64$$
$$x, y \geq 0.$$

(b)

$$\textbf{Minimize: } g(x,y) = 3x + 2y$$
$$\textbf{subject to: } 5x + 2y \geq 32$$
$$x + 3y \geq 22$$
$$x, y \geq 0.$$

(c)

$$\textbf{Minimize: } h(x,y,z) = -x - y$$
$$\textbf{subject to: } 3x + 6y + 2z \leq 6$$
$$y + z \geq 1$$
$$x, y, z \geq 0.$$

(d)

x_1	x_2	-1	
1	-1	3	$= -t_1$
-2	1	2	$= -t_2$
2	-1	0	$= f$

(e)

x_1	x_2	-1	
-1	-1	-2	$= -t_1$
-2	1	1	$= -t_2$
1	-2	0	$= -t_3$
2	-1	0	$= f$

(f)

y_1	-2	1	-3
y_2	1	-2	-2
1	0	0	7
	$= s_1$	$= s_2$	$= g$

6. For each problem in Exercise 2.6.5, sketch the feasible region and label the extreme points traversed by the Simplex Algorithm in order.

7. Solve the following using the Simplex Algorithm.

   (a) A firm produces a rare blend of scotch whiskey. The blend must contain at least 42% alcohol, at least 25% Highland blend, and no more than 15% malt. Three distillery products can be combined for the blend.

   Product A costs $12 a gallon, is 46% alcohol, 30% Highland blend and 10% malt. Product B costs $8 a gallon, is 40% alcohol, 20% Highland blend and 5% malt. Product C costs $14 a gallon, is 45% alcohol, 25% Highland blend and 2% malt.

   How much of each product should be used to produce 100 gallons of blend at minimal cost?

   (b) A company produces three types of tires for the SUV market. In their manufacture, the tires are processed on two machines, a molder and a capper. Tire type A takes 8 hours in the molder, 4 on the capper and sells for $45. Tire type B takes 10 hours in the molder, 7 on the capper and sells for $50. Tire type C takes 5 hours in the molder, 6 on the capper and sells for $40. At least 75 of each type of tire needs to be made each week to fulfill current contracts. If 3000 hours are available each week for molders and 2700 for cappers, how many of each type of tire should be made each week to maximize revenue?

8. The canonical optimization problem below potentially cycles (due to H.W. Kuhn). Solve the problem by using the Simplex Algorithm with Bland's Anticycling Theorem Theorem 2.3.5.

x_1	x_2	x_3	x_4	-1	
-2	-9	1	9	0	$= -t_1$
$1/3$	1	$-1/3$	-2	0	$= -t_2$
2	3	-1	-12	2	$= -t_3$
2	3	-1	-12	0	$= f$

9. Consider a tableau whose basic solution is feasible and optimal. Suppose each $b_i > 0$. Prove that this is the *unique* optimal solution if and only if each $c_j < 0$.

10. The following have multiple optimal solutions, use the Simplex Algorithm and then pivots to classify all the optimal solutions.

(a)

x_1	x_2	x_3	x_4	-1	
0	1	1	-1	3	$= -t_1$
1	1	1	-1	3	$= -t_2$
1	2	2	-4	0	$= f$

(b)

y_1	-1	-1	-1
y_2	-1	1	-1
-1	-2	1	0
	$= s_1$	$= s_2$	$= g$

11. Consider a square tableau:

	x_1	x_2	$\cdots$	x_n	-1	
	a_{11}	a_{12}	$\cdots$	a_{1n}	0	$= -t_1$
	a_{21}	a_{22}	$\cdots$	a_{2n}	0	$= -t_2$
	$\vdots$	$\vdots$	$\ddots$	$\vdots$	$\vdots$	$\vdots$
	a_{n1}	a_{n2}	$\cdots$	a_{nn}	0	$= -t_n$
	0	0	$\cdots$	0	0	

Suppose we perform pivots so we achieve a tableau of the form:

	t_1	t_2	$\cdots$	t_n	-1	
	a'_{11}	a'_{12}	$\cdots$	a'_{1n}	0	$= -x_1$
	a'_{21}	a'_{22}	$\cdots$	a'_{2n}	0	$= -x_2$
	$\vdots$	$\vdots$	$\ddots$	$\vdots$	$\vdots$	$\vdots$
	a'_{n1}	a'_{n2}	$\cdots$	a'_{nn}	0	$= -x_n$
	0	0	$\cdots$	0	0	

Let $A = [a_{ij}]_{n \times n}$ and $A' = [a'_{ij}]_{n \times n}$. For each of the following matrices A perform appropriate pivots to achieve A' and confirm $A' = A^{-1}$.

(a) $A = \begin{bmatrix} 0 & 1 & 1 \\ 1 & 1 & 0 \\ 1 & 0 & 1 \end{bmatrix}$.

(b) $A = \begin{bmatrix} 1 & 0 & 1 \\ 1 & 1 & 0 \\ 0 & 1 & -1 \end{bmatrix}$.

(c) $A = \begin{bmatrix} 0 & 1 & 2 & 3 \\ 1 & 2 & 3 & 2 \\ 2 & 3 & 2 & 1 \\ 3 & 2 & 1 & 0 \end{bmatrix}$.

12. Explain why in Exercise 2.6.11 that $A' = A^{-1}$ in general.

13.

(a) Find necessary and sufficient conditions for the minimization tableau

$$
\begin{array}{c|ccc|c|c}
y_1 & a_{11} & \cdots & a_{1m} & b_1 & \\
\vdots & \vdots & \ddots & \vdots & \vdots & \vdots \\
y_n & a_{n1} & \cdots & a_{nm} & b_n & \\
\hline
-1 & c_1 & \cdots & c_m & d & \\
& = s_1 & \cdots & = s_m & = g &
\end{array}
$$

to have a feasible basic solution.

(b) If a minimization tableau as depicted above has a feasible basic solution, must it also have a feasible basic maximum solution? Prove or find a counterexample.

(c) Find neccesary and sufficient conditions for

$$
\begin{array}{c|ccc|c|c}
& x_1 & \cdots & x_m & -1 & \\
\hline
y_1 & a_{11} & \cdots & a_{1m} & b_1 & = -t_1 \\
\vdots & \vdots & \ddots & \vdots & \vdots & \vdots \\
y_n & a_{n1} & \cdots & a_{nm} & b_n & = -t_n \\
\hline
-1 & c_1 & \cdots & c_m & d & = f \\
& = s_1 & \cdots & = s_m & = g &
\end{array}
$$

to encode feasible basic solutions for both its maximization and minimization problems.

14. Prove that each tableau always encodes a *unique* basic solution by first showing that the default starting basic solution is unique, and then proving that each pivot preserves the uniqueness of the basic solution.

   Hint. Note that each basic solution is the intersection of m hyperplanes. What would it take for this to be empty or contain multiple points? Think in terms of linear (in)dependence and solving linear systems.

Chapter 3

Noncanonical Problems

Thus far, we have focused on *canonical* optimization problems where each decision variable is nonnegative, and all bounds are inequalities. However, what would happen if we relaxed, or strengthened these conditions?

In this chapter, we explore noncanonical optimization problems. In Section 3.1 we consider the possibility of potentially negative decision variables, and explore the algebraic and geometric interpretations of this phenomena. In Section 3.2 we do the same for bounds defined by equality conditions, rather than inequalities. In Section 3.3 we show how to encode and solve noncanonical linear optimization using Sage.

3.1 Unconstrained Variables

Recall that canonical problems have nonnegative variables and inequality bounds. In this section, we consider linear optimization problems with potentially negative variables.

Activity 3.1.1 Suppose we wanted to solve the linear optimization problem:

$$\textbf{Maximize: } f(\mathbf{x}) = 3x + 2y$$
$$\textbf{subject to: } 4x + 5y \leq 23$$
$$2x + y \leq 7$$
$$x - y \leq 5.$$

(a) What are some differences between this linear optimization problem and previous examples of optimization problems?

(b) We can record the problem with the following tableau, we denote the variables which can be negative by circling them:

$\textcircled{x}$	$\textcircled{y}$	-1	
4	5	23	$= -t_1$
2	1	7	$= -t_2$
1	-1	5	$= -t_3$
3	2	0	$= f$

What point currently represents the basic solution?

(c) Suppose we pivot on the entry with the $*$. What point has the basic solution moved to? (You do not need to fill in the tableau.)

x	y	-1	
4	5	23	$= -t_1$
2*	1	7	$= -t_2$
1	-1	5	$= -t_3$
3	2	0	$= f$

(d) Suppose we pivot on the entry with the $*$. What point has the basic solution moved to?

t_2	y	-1	
?	?	?	$= -t_1$
?	?	?	$= -x$
?	?*	?	$= -t_3$
?	?	?	$= f$

(e) Consider the plot of the feasible region. What geometrically did our two pivots do?

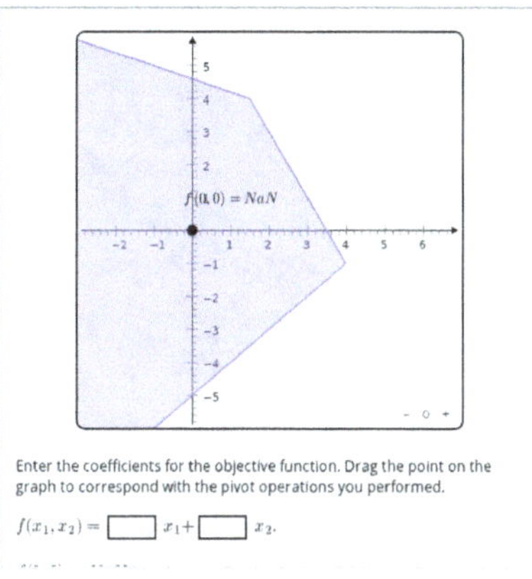

(f) List all the hyperplanes that bound the feasible region.

(g) Recall that each simplex pivot represents a shift of the basic solution from one extreme point to another. Which of the following do you believe is true?

    A. Following the rules of pivoting through the Simplex Algorithm, we should be able to return to the origin, and this is consistent with our geometric viewpoint.

    B. It is possible to perform pivots that don't necessarily follow the rules of the Simplex Algorithm, to return to the origin, and this is consistent with our geometric viewpoint.

    C. It is technically possible to perform pivots that don't necessarily follow the rules of the Simplex Algorithm to return to the origin, and this is *not* consistent with our geometric viewpoint.

Definition 3.1.2 In a linear optimization problem, a variable which can be potentially negative is called an **unconstrained** variable. ◇

Activity 3.1.3 Suppose we wanted to solve the linear optimization problem:

$$\text{Maximize: } f(\mathbf{x}) = 5x + 4y$$
$$\text{subject to: } 2x + 3y \leq 26$$
$$-2x - 10y \leq 2.$$

(a) We can record the problem as follows:

x	y	-1	
2	3	26	$= -t_1$
-2	-10	2	$= -t_2$
5	4	0	$= f$

Out of x, y, t_1, t_2, which are nonnegative?

(b) Perform a pivot on the entry with the $*$:

$\textcircled{x}$	$\textcircled{y}$	-1	
2^*	3	26	$= -t_1$
-2	-10	2	$= -t_2$
5	4	0	$= f$

(c) Consider the resulting tableau:

t_1	$\textcircled{y}$	-1	
?	?	?	$= -\textcircled{x}$
?	?	?	$= -t_2$
?	?	?	$= f$

What point in $\mathbb{R}^2$ represents the basic solution of this tableau? Why is this point *not* an optimal solution?

(d) Consider the row with the $\textcircled{x}$, and recall that $\textcircled{x}$ is allowed to be negative. Consider only this row and the nonnegativity constraints of t_i. Which of the following is t_1 allowed to be?

A. $t_1 = 0$.   B. $t_1 = 10$.   C. $t_1 = 100$.

For each choice of t_1 that is valid, with $y = 0$, what would x be?

(e) Which of the following bits of information does this row communicate to us? (There could be more than one.)

A. There is a lower bound for t_1 which is greater than 0.

B. There is an upper bound for t_1.

C. There is a lower bound for t_2 which is greater than 0.

D. There is an upper bound for t_2.

E. There is a lower bound for x which is greater than 0.

F. There is an upper bound for x.

G. There is a lower bound for y which is greater than 0.

H. There is an upper bound for y.

I. That $x = 13 - \frac{1}{2}t_1 - \frac{3}{2}y$.

J. That $t_2 = 28 - t_1 + 7y$.

(f) After recording this piece of information, we may as well delete this row, since we have what we need from it:

t_1	$ⓨ$	-1	
?	?*	?	$= -t_2$
?	?	?	$= f$

Pivot on the entry with the $*$.

(g) Why does the resulting tableau encode a basic solution which is *not* infeasible?

(h) Which of the following bits of information does the $ⓨ$ row communicate to us? (There could be more than one.)

A. There is a lower bound for t_1 which is greater than 0.

B. There is an upper bound for t_1.

C. There is a lower bound for t_2 which is greater than 0.

D. There is an upper bound for t_2.

E. There is a lower bound for y which is greater than 0.

F. There is an upper bound for y.

G. That $y = -\frac{1}{7}t_1 + \frac{1}{7}t_2 - 4$.

(i) After recording this piece of information, we may as well delete this row, since we have what we need from it:

t_1	t_2	-1	
?	?	?	$= -f$

Why is the basic solution encoded by this tableau optimal?

(j) What are the final values for t_1, t_2, x, y, f?

(k) Consider the plot of the feasible region. If we started at the origin, what did we do in each step?

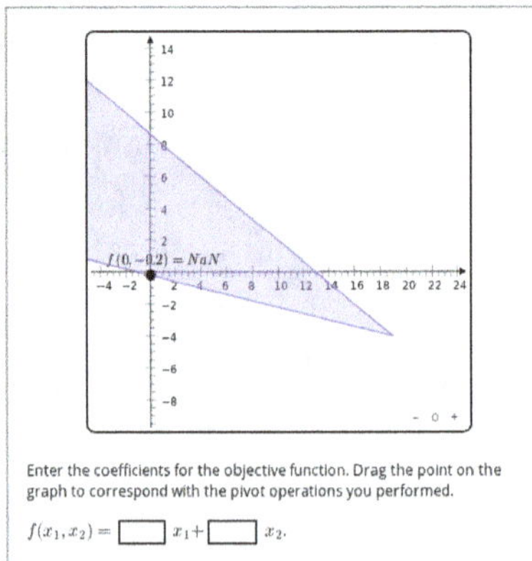

Observation 3.1.4 With unconstrained variables, we proceed as follows.

1. Remove all unconstrained slack variables and delete the corresponding rows.

2. If there are no unconstrained decision variables: STOP.

3. Pivot an unconstrained decision variable with a slack variable.

4. GOTO 1.

Activity 3.1.5 Solve the linear optimization problem:

$$\textbf{Maximize: } f(\mathbf{x}) = x + 3y$$
$$\textbf{subject to: } x + 2y \le 10$$
$$2x + y \le 15$$
$$x \ge 0.$$

3.2 Super Constrained Bounds

In this section, we consider linear optimization problems with equality bounds.

Activity 3.2.1 Suppose we wanted to solve the linear optimization problem:

$$\textbf{Maximize: } f(\mathbf{x}) = x + y + z$$
$$\textbf{subject to: } 2x + 3z \leq 2$$
$$4x - y \leq 1$$
$$5y - 2z \leq 5$$
$$x, y, z \geq 0$$

(a) Plot the feasible region, what dimension is it?

   Hint.

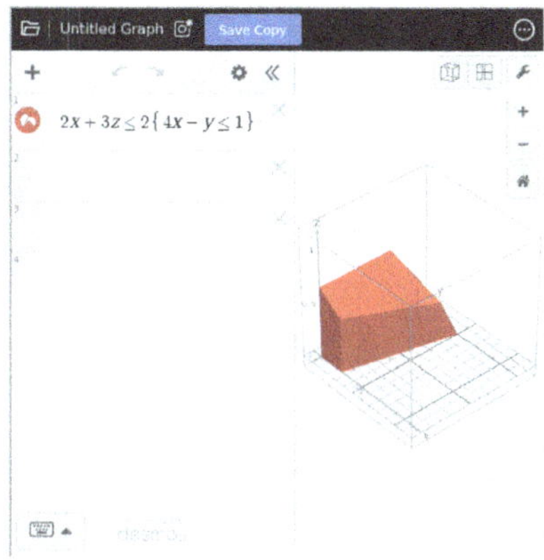 

Standalone

(b) Suppose we added in the constraint $x + y - 2z = 1$ Plot the feasible region, what dimension is it?

(c) Consider the inequality $x + y - 2z \leq 1$ captured by the equality $x + y - 2z - 1 = -t_4$. What value must t_4 so that $x + y - 2z \leq 1$ is always an equality? Call this value ξ.

   Hint.

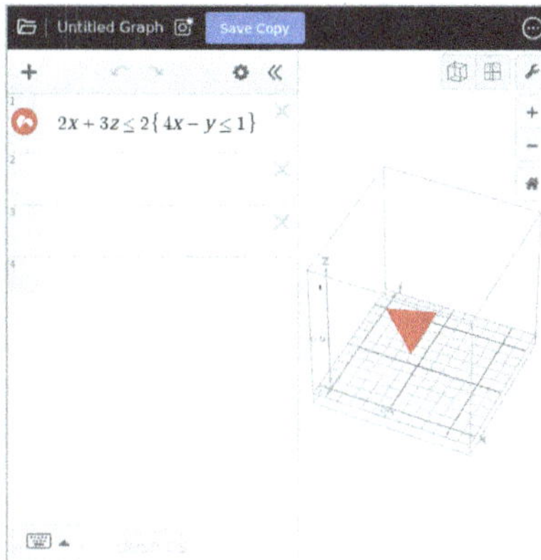 

Standalone

(d) Note that this problem may be encoded in the tableau:

$$
\begin{array}{cccc}
x & y & z & -1 \\
\end{array}
$$

x	y	z	-1	
2	0	3	2	$= -t_1$
4	-1	0	1	$= -t_2$
0	5	-2	5	$= -t_3$
1	1^*	-2	1	$= -\xi$
1	1	1	0	$= f$

Without computing the tableau, what point does the basic solution move to if we pivot on the entry with a $*$? Is it feasible?

(e) As we traverse corner points on the way to an optimal solution, would we ever leave the plane $x + y - 2z = 1$?

(f) After the last pivot, our tableau has the form:

x	ξ	z	-1	
a_{11}	a_{12}	a_{13}	b_1	$= -t_1$
a_{21}	a_{22}	a_{23}	b_2	$= -t_2$
a_{31}	a_{32}	a_{33}	b_3	$= -t_3$
a_{41}	a_{42}	a_{43}	a_{44}	$= -y$
c_1	c_2	c_3	d	$= f$

Ignoring the specific values of the entries of this tableau, using the value for ξ computed earlier, rewrite each of the above equations in terms of x, z. What information did the ξ column provide?

Observation 3.2.2 An equality constraint can be written as:

$$
a_{i1}x_1 + a_{i2}x_2 + \cdots + a_{im}x_m - b_i = 0
$$

and thus recorded in a tableau as:

x_1	x_2	$\cdots$	x_{j-1}	x_j	x_{j+1}	$\cdots$	x_m	-1	
a_{11}	a_{12}	$\cdots$	$a_{1(j-1)}$	a_{1j}	$a_{1(j+1)}$	$\cdots$	a_{1m}	b_1	$= -t_1$
a_{21}	a_{22}	$\cdots$	$a_{2(j-1)}$	a_{2j}	$a_{2(j+1)}$	$\cdots$	a_{2m}	b_2	$= -t_2$
$\vdots$	$\vdots$	$\ddots$	$\vdots$	$\vdots$	$\vdots$	$\ddots$	$\vdots$	$\vdots$	$\vdots$
a_{i1}	a_{i2}	$\cdots$	$a_{i(j-1)}$	a_{ij}	$a_{i(j+1)}$	$\cdots$	a_{im}	b_i	$= -0$
$\vdots$	$\vdots$	$\ddots$	$\vdots$	$\vdots$	$\vdots$	$\ddots$	$\vdots$	$\vdots$	$\vdots$
a_{n1}	a_{n2}	$\cdots$	$a_{n(j-1)}$	a_{nj}	$a_{n(j+1)}$	$\cdots$	a_{nm}	b_n	$= -t_n$
c_1	c_2	$\cdots$	c_{j-1}	c_j	c_{j+1}	$\cdots$	c_m	d	

If we pivot on the a_{ij}^* entry, we obtain:

x_1	x_2	$\cdots$	x_{j-1}	0	x_{j+1}	$\cdots$	x_m	-1	
a_{11}	a_{12}	$\cdots$	$a_{1(j-1)}$	a_{1j}	$a_{1(j+1)}$	$\cdots$	a_{1m}	b_1	$=-t_1$
a_{21}	a_{22}	$\cdots$	$a_{2(j-1)}$	a_{2j}	$a_{2(j+1)}$	$\cdots$	a_{2m}	b_2	$=-t_2$
$\vdots$	$\vdots$	$\ddots$	$\vdots$	$\vdots$	$\vdots$	$\ddots$	$\vdots$	$\vdots$	$\vdots$
a_{i1}	a_{i2}	$\cdots$	$a_{i(j-1)}$	a_{ij}	$a_{i(j+1)}$	$\cdots$	a_{im}	b_i	$=-x_j$
$\vdots$	$\vdots$	$\ddots$	$\vdots$	$\vdots$	$\vdots$	$\ddots$	$\vdots$	$\vdots$	$\vdots$
a_{n1}	a_{n2}	$\cdots$	$a_{n(j-1)}$	a_{nj}	$a_{n(j+1)}$	$\cdots$	a_{nm}	b_n	$=-t_n$
c_1	c_2	$\cdots$	c_{j-1}	c_j	c_{j+1}	$\cdots$	c_m	d	

Then depending on your perspective, we can either delete the 0 column because it does not contribute information algebraically, or because it is redundant geometrically, and we restrict ourselves to a $m-1$ dimensional solution space. Either way, removing this column gives us:

x_1	x_2	$\cdots$	x_{j-1}	x_{j+1}	$\cdots$	x_m	-1	
a_{11}	a_{12}	$\cdots$	$a_{1(j-1)}$	$a_{1(j+1)}$	$\cdots$	a_{1m}	b_1	$=-t_1$
a_{21}	a_{22}	$\cdots$	$a_{2(j-1)}$	$a_{2(j+1)}$	$\cdots$	a_{2m}	b_2	$=-t_2$
$\vdots$	$\vdots$	$\ddots$	$\vdots$	$\vdots$	$\ddots$	$\vdots$	$\vdots$	$\vdots$
a_{i1}	a_{i2}	$\cdots$	$a_{i(j-1)}$	$a_{i(j+1)}$	$\cdots$	a_{im}	b_i	$=-x_j$
$\vdots$	$\vdots$	$\ddots$	$\vdots$	$\vdots$	$\ddots$	$\vdots$	$\vdots$	$\vdots$
a_{n1}	a_{n2}	$\cdots$	$a_{n(j-1)}$	$a_{n(j+1)}$	$\cdots$	a_{nm}	b_n	$=-t_n$
c_1	c_2	$\cdots$	c_{j-1}	c_{j+1}	$\cdots$	c_m	d	

Activity 3.2.3 Consider the linear optimization problem:

$$\textbf{Maximize: } f(\mathbf{x}) = 2x + y - 2z$$
$$\textbf{subject to: } x + y + z \leq 1$$
$$y + 4z = 2$$
$$x, y, z \geq 0.$$

(a) Record this problem in a tableau with an equality constraint.

x	y	z	-1	
a_{11}	a_{12}	a_{13}	b_1	$=-t_1$
a_{21}	a_{22}	a_{23}	b_2	$=-0$
c_1	c_2	c_3	d	

(b) Pivot on the entry with $*$.

x	y	z	-1	
a_{11}	a_{12}	a_{13}	b_1	$=-t_1$
a_{21}	a_{22}^{*}	a_{23}	b_2	$=-0$
c_1	c_2	c_3	d	

(c) We obtained a tableau of the form:

x	0	z	-1	
a_{11}	a_{12}	a_{13}	b_1	$= -t_1$
a_{21}	a_{22}	a_{23}	b_2	$= -y$
c_1	c_2	c_3	d	

Rewrite the 3 rows as linear equalities, and verify that the 0 column contributes nothing.

(d) Delete the 0 column and solve the remaining system.

Hint.

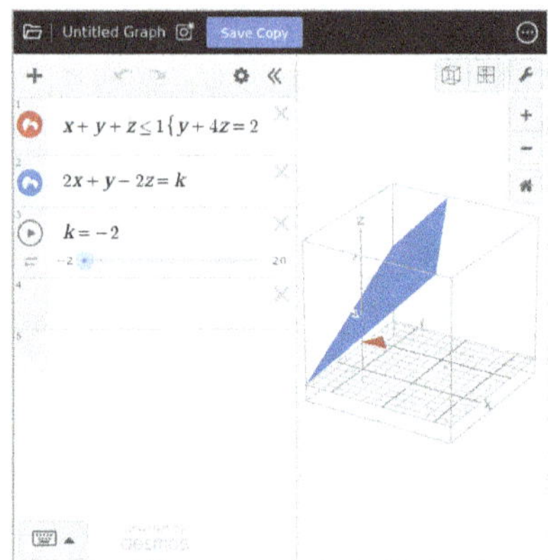

Standalone

Activity 3.2.4 Solve the linear optimization problem:

$$\textbf{Maximize:} \ \ f(\mathbf{x}) = x + 4y + 2z$$
$$\textbf{subject to:} \ \ x + 2y + 3z \le 6$$
$$4x - 7y = 28$$
$$x, y, z \ge 0.$$

Hint.

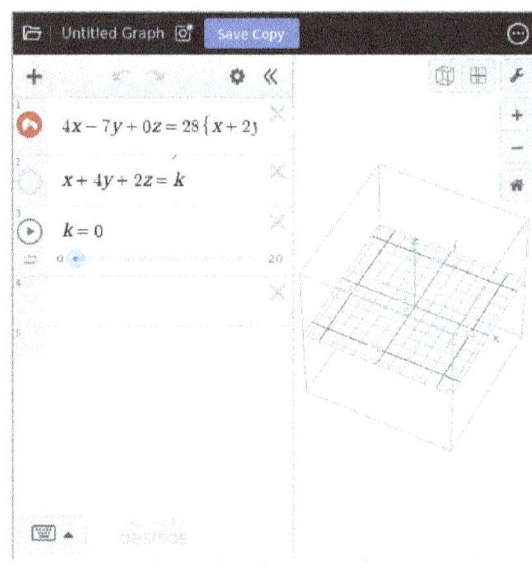

Standalone

Activity 3.2.5 Solve the linear optimization problem:

$$\textbf{Maximize:} \ \ f(\mathbf{x}) = -x + y + 2z$$
$$\textbf{subject to:} \ \ x + 4y - 2z = 6$$
$$2x - y \le 10$$

$$y + z \leq 5$$
$$x, y, z \geq 0.$$

Hint.

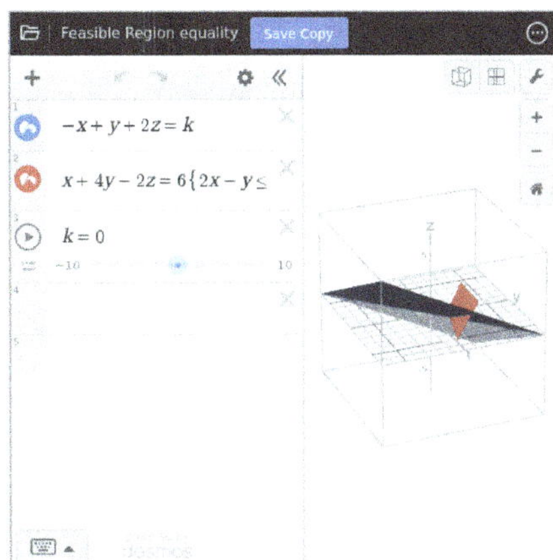

Standalone

Activity 3.2.6 Solve the linear optimization problem:

$$\textbf{Maximize: } f(\mathbf{x}) = x + 2y + z$$
$$\textbf{subject to: } x + y + z = 6$$
$$x + y \leq 1$$
$$x, z \geq 0.$$

Hint.

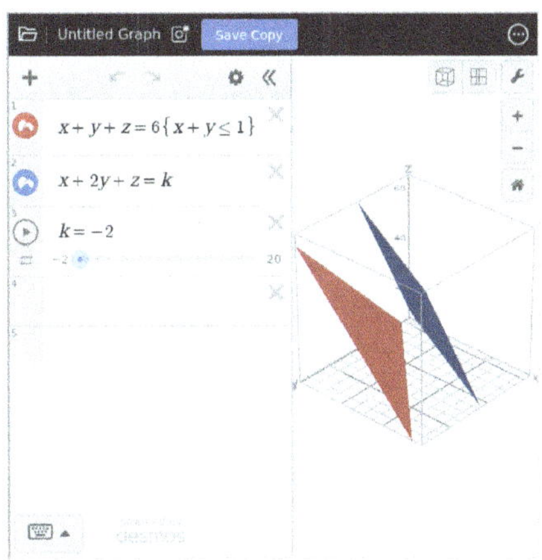

Standalone

3.3 Solving Noncanonical Problems with Sage

In Section 2.4, we showed how to use Sage to solve canonical linear optimization problems with the Simplex Algorithm. In this section, we use Sage to solve noncanonical problems.

Activity 3.3.1 Say we want to solve the noncanonical linear optimization problem:

$$\textbf{Minimize: } f(\mathbf{x}) = 3x + y + 2z$$
$$\textbf{subject to: } x + 2y + 3z \geq 24$$
$$2x + 4y + 3z = 36$$
$$y, z \geq 0.$$

(a) Record this noncanonical problem using Sage:

```
%display typeset
A = ([[1,2,3],[2,4,3]])
b = ([24,36])
c = ([3,1,2])
P = InteractiveLPProblem(A, b, c,
  ["x", "y", "z"],
  constraint_type = [">=", "=="],
  variable_type = ["", ">=", ">="],
  problem_type = "min")
P
```

(b) Find the optimal solution:

```
print(P.optimal_solution())
print(P.optimal_value())
```

Note that we use the command `InteractiveLPProblem` for general (potentially noncanonical) linear optimization problems, rather than `InteractiveLPProblemStandardForm`. Sage does not have a command for the Simplex Algorithm for `InteractiveLPProblem`.

Activity 3.3.2 Solve:

$$\textbf{Minimize: } f(\mathbf{x}) = -5x + y - 2z$$
$$\textbf{subject to: } 2x + z = 0$$
$$x - y \geq 1$$
$$3x - y + z \leq 3.$$

```
%display typeset
A = (FIXME)
b = (FIXME)
c = (FIXME)
P = InteractiveLPProblem(A, b, c,
  [FIXME],
  constraint_type = [FIXME],
  variable_type = [FIXME],
  problem_type = FIXME)
P
```

```
print(P.optimal_solution())
print(P.optimal_value())
```

3.4 Summary of Chapter 3

We first consider the case where some x_j is allowed to be negative, which we denote as $\boxed{x_j}$.

x_1	x_2	$\cdots$	x_{j-1}	$\boxed{x_j}$	x_{j+1}	$\cdots$	x_m	-1	
a_{11}	a_{12}	$\cdots$	$a_{1(j-1)}$	a_{1j}	$a_{1(j+1)}$	$\cdots$	a_{1m}	b_1	$= -t_1$
a_{21}	a_{22}	$\cdots$	$a_{2(j-1)}$	a_{2j}	$a_{2(j+1)}$	$\cdots$	a_{2m}	b_2	$= -t_2$
$\vdots$	$\vdots$	$\ddots$	$\vdots$	$\vdots$	$\vdots$	$\ddots$	$\vdots$	$\vdots$	$\vdots$
a_{i1}	a_{i2}	$\cdots$	$a_{i(j-1)}$	a_{ij}	$a_{i(j+1)}$	$\cdots$	a_{im}	b_i	$= -t_i$
$\vdots$	$\vdots$	$\ddots$	$\vdots$	$\vdots$	$\vdots$	$\ddots$	$\vdots$	$\vdots$	$\vdots$
a_{n1}	a_{n2}	$\cdots$	$a_{n(j-1)}$	a_{nj}	$a_{n(j+1)}$	$\cdots$	a_{nm}	b_n	$= -t_n$
c_1	c_2	$\cdots$	c_{j-1}	c_j	c_{j+1}	$\cdots$	c_m	d	

Note that should we pivot on a_{ij}

x_1	x_2	$\cdots$	x_{j-1}	t_i	x_{j+1}	$\cdots$	x_m	-1	
a_{11}	a_{12}	$\cdots$	$a_{1(j-1)}$	a_{1j}	$a_{1(j+1)}$	$\cdots$	a_{1m}	b_1	$= -t_1$
a_{21}	a_{22}	$\cdots$	$a_{2(j-1)}$	a_{2j}	$a_{2(j+1)}$	$\cdots$	a_{2m}	b_2	$= -t_2$
$\vdots$	$\vdots$	$\ddots$	$\vdots$	$\vdots$	$\vdots$	$\ddots$	$\vdots$	$\vdots$	$\vdots$
a_{i1}	a_{i2}	$\cdots$	$a_{i(j-1)}$	a_{ij}	$a_{i(j+1)}$	$\cdots$	a_{im}	b_i	$= -\boxed{x_j}$
$\vdots$	$\vdots$	$\ddots$	$\vdots$	$\vdots$	$\vdots$	$\ddots$	$\vdots$	$\vdots$	$\vdots$
a_{n1}	a_{n2}	$\cdots$	$a_{n(j-1)}$	a_{nj}	$a_{n(j+1)}$	$\cdots$	a_{nm}	b_n	$= -t_n$
c_1	c_2	$\cdots$	c_{j-1}	c_j	c_{j+1}	$\cdots$	c_m	d	

that having a slack variable that can be negative serves no purpose, since the point of the nonnegativity constraint is to reinforce the inequality. Also note that the hyperplane $x_j = 0$ also plays no role in this problem, since it does not serve as a bounding hyperplane. Thus, we may simply delete this row (after recording the equality) as this equality will have no bearing on future steps or the final solution.

x_1	x_2	$\cdots$	x_{j-1}	t_i	x_{j+1}	$\cdots$	x_m	-1	
a_{11}	a_{12}	$\cdots$	$a_{1(j-1)}$	a_{1j}	$a_{1(j+1)}$	$\cdots$	a_{1m}	b_1	$= -t_1$
a_{21}	a_{22}	$\cdots$	$a_{2(j-1)}$	a_{2j}	$a_{2(j+1)}$	$\cdots$	a_{2m}	b_2	$= -t_2$
$\vdots$	$\vdots$	$\ddots$	$\vdots$	$\vdots$	$\vdots$	$\ddots$	$\vdots$	$\vdots$	$\vdots$
a_{n1}	a_{n2}	$\cdots$	$a_{n(j-1)}$	a_{nj}	$a_{n(j+1)}$	$\cdots$	a_{nm}	b_n	$= -t_n$
c_1	c_2	$\cdots$	c_{j-1}	c_j	c_{j+1}	$\cdots$	c_m	d	

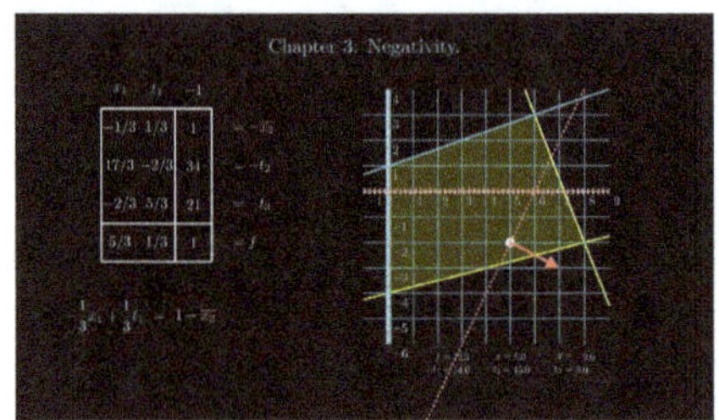

Figure 3.4.1 A summary of unconstrained variables.

Next we consider the case where the feasible region is bound by an equality, rather than an inequality. In such a case, the slack variable t_i must be zero.

$$
\begin{array}{|cccccccc|cl}
x_1 & x_2 & \cdots & x_{j-1} & x_j & x_{j+1} & \cdots & x_m & -1 & \\
\hline
a_{11} & a_{12} & \cdots & a_{1(j-1)} & a_{1j} & a_{1(j+1)} & \cdots & a_{1m} & b_1 & = -t_1 \\
a_{21} & a_{22} & \cdots & a_{2(j-1)} & a_{2j} & a_{2(j+1)} & \cdots & a_{2m} & b_2 & = -t_2 \\
\vdots & \vdots & \ddots & \vdots & \vdots & \vdots & \ddots & \vdots & \vdots & \vdots \\
a_{i1} & a_{i2} & \cdots & a_{i(j-1)} & a_{ij} & a_{i(j+1)} & \cdots & a_{im} & b_i & = -0 \\
\vdots & \vdots & \ddots & \vdots & \vdots & \vdots & \ddots & \vdots & \vdots & \vdots \\
a_{n1} & a_{n2} & \cdots & a_{n(j-1)} & a_{nj} & a_{n(j+1)} & \cdots & a_{nm} & b_n & = -t_n \\
\hline
c_1 & c_2 & \cdots & c_{j-1} & c_j & c_{j+1} & \cdots & c_m & d &
\end{array}
$$

Now in this case, if we pivot on a_{ij} entry, we obtain:

$$
\begin{array}{|cccccccc|cl}
x_1 & x_2 & \cdots & x_{j-1} & 0 & x_{j+1} & \cdots & x_m & -1 & \\
\hline
a_{11} & a_{12} & \cdots & a_{1(j-1)} & a_{1j} & a_{1(j+1)} & \cdots & a_{1m} & b_1 & = -t_1 \\
a_{21} & a_{22} & \cdots & a_{2(j-1)} & a_{2j} & a_{2(j+1)} & \cdots & a_{2m} & b_2 & = -t_2 \\
\vdots & \vdots & \ddots & \vdots & \vdots & \vdots & \ddots & \vdots & \vdots & \vdots \\
a_{i1} & a_{i2} & \cdots & a_{i(j-1)} & a_{ij} & a_{i(j+1)} & \cdots & a_{im} & b_i & = -x_j \\
\vdots & \vdots & \ddots & \vdots & \vdots & \vdots & \ddots & \vdots & \vdots & \vdots \\
a_{n1} & a_{n2} & \cdots & a_{n(j-1)} & a_{nj} & a_{n(j+1)} & \cdots & a_{nm} & b_n & = -t_n \\
\hline
c_1 & c_2 & \cdots & c_{j-1} & c_j & c_{j+1} & \cdots & c_m & d &
\end{array}
$$

Note that what the coefficients a_{ij}, c_j are, they are coefficients of the "variable" 0 and are thus irrelevant. Moreover, we cannot pivot away from this plane, since our feasible region is contained completely within the plane. Thus we may delete this column.

$$
\begin{array}{|ccccccc|cl}
x_1 & x_2 & \cdots & x_{j-1} & x_{j+1} & \cdots & x_m & -1 & \\
\hline
a_{11} & a_{12} & \cdots & a_{1(j-1)} & a_{1(j+1)} & \cdots & a_{1m} & b_1 & = -t_1 \\
a_{21} & a_{22} & \cdots & a_{2(j-1)} & a_{2(j+1)} & \cdots & a_{2m} & b_2 & = -t_2 \\
\vdots & \vdots & \ddots & \vdots & \vdots & \ddots & \vdots & \vdots & \vdots \\
a_{i1} & a_{i2} & \cdots & a_{i(j-1)} & a_{i(j+1)} & \cdots & a_{im} & b_i & = -x_j \\
\vdots & \vdots & \ddots & \vdots & \vdots & \vdots & \ddots & \vdots & \vdots \\
a_{n1} & a_{n2} & \cdots & a_{n(j-1)} & a_{n(j+1)} & \cdots & a_{nm} & b_n & = -t_n \\
\hline
c_1 & c_2 & \cdots & c_{j-1} & c_{j+1} & \cdots & c_m & d &
\end{array}
$$

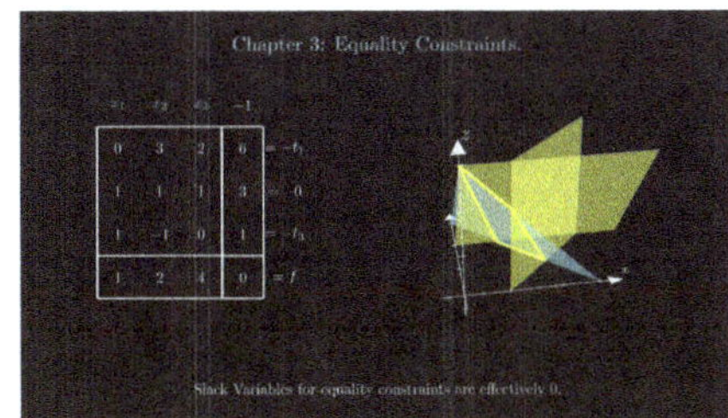

Figure 3.4.2 A summary of equality constraints.

3.5 Problems for Chapter 3

Exercises

1. Solve each of the noncanonical linear optimization problems below. If a linear optimization problem has infinitely many optimal solutions, find all optimal solutions.

   (a)

   $$\textbf{Maximize: } f(x, y, z) = x - 2y - z$$
   $$\textbf{subject to: } x + y \leq 1$$
   $$x + y + z \geq 2$$
   $$y - 2z \geq 0.$$

   (b)

   $$\textbf{Maximize: } g(x, y, z) = 4x + 2y + z$$
   $$\textbf{subject to: } x + y + z \leq 5$$
   $$4x - y - z \leq 5$$
   $$2y + z \leq 4.$$

   (c)

   $$\textbf{Maximize: } f(x, y, z) = 4x + 2y + z$$
   $$\textbf{subject to: } x + z \leq 5$$
   $$4x - y - z \leq 5$$
   $$2y + z \leq 4.$$

   (d)

   $$\textbf{Maximize: } k(x, y, z) = x + 2y + z$$
   $$\textbf{subject to: } 2x - y + z = 3$$
   $$x + 2y \leq 4$$
   $$-2x + 5y + 3z \leq 4$$
   $$x, y, z \geq 0.$$

   (e)

   $$\textbf{Minimize: } h(x, y, z) = x + y + z$$
   $$\textbf{subject to: } x - y \geq 1$$
   $$2x + z \leq 10$$
   $$y + z = 4$$
   $$x, y, z \geq 0.$$

   (f)

   $$\textbf{Minimize: } f(x, y, z) = 3x + y + 2z$$
   $$\textbf{subject to: } x + 2y + 3z \geq 24$$
   $$2x + 4y + 3z = 36$$
   $$y, z \geq 0.$$

(g)

$$\textbf{Maximize: } \ell(x, y, z) = x + y + z$$
$$\textbf{subject to: } x + y + z \leq 3$$
$$x + y \leq 1$$
$$y + 2z = 2$$
$$x, y \geq 0.$$

(h)

$$\textbf{Maximize: } f(x, y, z) = 3x - 2y + 3z$$
$$\textbf{subject to: } x - y + 2z = 6$$
$$x + 2z = 8$$
$$y + 2z \geq 2$$
$$y, z \geq 0.$$

(i)

$$\textbf{Minimize: } \alpha(x, y, z) = -5x + y - 2z$$
$$\textbf{subject to: } 2x + z = 0$$
$$x - y \geq 1$$
$$3x - y + z \leq 3.$$

(j)

$$\textbf{Maximize: } g(x, y, z) = x + y + z$$
$$\textbf{subject to: } x - y - z \leq 2.$$
$$y - z \geq 1$$

(k)

$$\textbf{Maximize: } f(x, y) = x + y$$
$$\textbf{subject to: } 2x + y = 5$$
$$x - y = -2$$
$$x + 2y = 8$$
$$x, y \geq 0.$$

2. Label each of the following statements TRUE or FALSE. If the statement is FALSE, provide a counterexample.

    (a) A noncanonical linear optimization problem with more unconstrained independent variables than constraints has unbounded objective function (as in Exercise 3.5.1 **(j)**).

    (b) A noncanonical linear optimization problem with more equations of constraint than independent variables is infeasible (as in Exercise 3.5.1 **(k)**).

3. Sketch the constraint set for each noncanonical linear optimization problem below. On the basis of this constraint set, formulate a conjecture as to whether or not the solution of the given problem is the same as the solution of the associated canonical linear optimization problem where all independent variables are constrained to be nonnegative. Verify your conjecture by solving both linear optimization problems.

(a)

$$\textbf{Maximize: } f(x,y) = x + y$$
$$\textbf{subject to: } 2x + y \leq 11$$
$$x - y \geq -2.$$

(b)

$$\textbf{Minimize: } g(x,y) = 2x + y$$
$$\textbf{subject to: } 3x + 2y \geq 5$$
$$2x - y \geq 1.$$

(c)

$$\textbf{Maximize: } f(x,y) = 5x + y$$
$$\textbf{subject to: } x - y \leq 8$$
$$2x + y \leq 7.$$

(d)

$$\textbf{Minimize: } g(x,y) = x + 2y$$
$$\textbf{subject to: } x + 3y \geq 5$$
$$2x + y \geq 0.$$

4. The following problem has infinitely many optimal solutions. Sketch the feasible region, find the optimal solutions and sketch the solution set.

$$\textbf{Maximize: } f(x,y,z) = 2x + y - 2z$$
$$\textbf{subject to: } x + y + z \leq 1$$
$$y + 4z = 2$$
$$x, y, z \geq 0.$$

5. Another method method for transforming a linear optimization problem with unconstrained independent variables into canonical form is to replace every unconstrained independent variable by the difference of two independent variables constrained to be nonnegative. This produces an equivalent canonical linear optimization problem which is solved by using the Simplex Algorithm.

For example:

$$\textbf{Maximize: } f(x,y) = 4x + y$$
$$\textbf{subject to: } x + y \leq 2$$
$$x - 2y \leq 5$$
$$x \geq 0$$

may be restated as the following canonical optimization problem:

$$\textbf{Maximize: } f(x, y^+, y^-) = 4x + y^+ - y^-$$

$$\textbf{subject to: } x + y^+ - y^- \le 2$$
$$x - 2y^+ + 2y^- \le 5$$
$$x, y^+, y^- \ge 0$$

where we let $y = y^+ - y^-$.

(a) Solve the second canonical optimization problem above.

(b) Solve the original optimization problem.

(c) How do the solutions compare?

6. Another method method for transforming a linear optimization problem with equations of constraint into canonical form is to replace every equation of constraint by two inequality constraints. This produces an equivalent canonical linear optimization problem which is solved by using the Simplex Algorithm.

For example:

$$\textbf{Maximize: } f(x, y, z) = x + y + z$$
$$\textbf{subject to: } x + 2z \le 7$$
$$-x + 2y - z \le 3$$
$$x + y = 7$$
$$x, y, z \ge 0$$

may be restated as the following canonical optimization problem:

$$\textbf{Maximize: } f(x, y, z) = x + y + z$$
$$\textbf{subject to: } x + 2z \le 7$$
$$-x + 2y - z \le 3$$
$$x + y \le 7$$
$$x + y \ge 7$$
$$x, y, z \ge 0.$$

(a) Solve the second canonical optimization problem above.

(b) Solve the original optimization problem.

(c) How do the solutions compare?

7. Use the methods presented in Exercise 3.5.5 and Exercise 3.5.6 to solve:

$$\textbf{Maximize: } f(x, y, z) = 2x + y - 2z$$
$$\textbf{subject to: } 2x + y \le 3$$
$$x + 2y - z \le 5$$
$$x + y + z = 0$$
$$x, y \ge 0.$$

8. What are some pros and cons of the methods presented in Exercise 3.5.5 and Exercise 3.5.6?

Chapter 4

Duality

One of the fundamental ideas of linear optimization is that of duality. We saw a precursor to this idea in both Activity 2.1.6 and Activity 2.1.12. The same initial scenario produces both a maximization and a minimization problem. What does this mean, why does this occur, and what are the relations between these problems?

In this chapter, we define and explore the notion of duality in linear optimization. In Section 4.1 we give a geometric and algebraic interpretation of dual variables and the dual problem. In Section 4.2 we use geometric reasoning to guide proofs of powerful duality results connecting the primal and dual problems. Finally, in Section 4.3 we show how the simplex pivot from Chapter 2 applies geometrically and computationally to the dual problem, and show how Tucker tableaus encode the dual problem.

4.1 Sensitivity Analysis

In this section, we begin the exploration of what duality means. We assign natural meanings to dual variables and the dual problem. One perspective to keep in mind this section is the role of bounds and objective function in the primal problem, and how they change here.

Exploration 4.1.1 The witch Agnesi[1] is brewing a healing elixir and a poison. A pint of healing elixir takes 3 newt eyes and one frog, whereas a pint of poison takes 1 each of newt eyes and frogs. She currently has 34 newt eyes and 14 frogs.

Supposing that the healing elixir sells for three gold pieces, and the poison sells for two. Agnesi wishes to maximize her revenue. Let us also suppose that since these are liquids, she is happy making fractional amounts of elixirs and potions.

(a) Before proceeding to solve the problem, make an estimate: how much do you think each newt eye and frog is worth to her? Why do you think so?

(b) We now return to the initially posed maximization problem. Sketch the feasible region for this problem, and use whatever method you feel like to find the optimal solution.

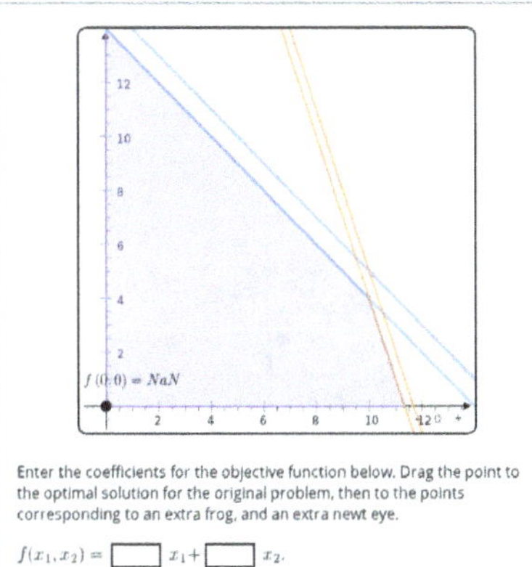

$f(x_1, x_2) = \boxed{} \; x_1 + \boxed{} \; x_2.$

(c) Agnesi is frustrated by her production levels and income. She is going to recruit some local children to gather more materials for her. Without explicitly computing anything, looking at her situation, what would result in greater profits for her, more newt eyes or frogs?

(d) Recompute this linear optimization problem with 35 newt eyes and 14 frogs, then with 34 newt eyes and 15 frogs. Which provides the greater increase in revenue? Is this consistent with what you thought earlier? (Use the above interactive.)

(e) If the need for healing elixir increases and they now sell for 5 gold, would that change our answers above?

Activity 4.1.2 In both Exploration 4.1.1 and Activity 2.1.12 we essentially explore the idea of assigning values somehow to the bounds of a maximization problem.

Suppose you have a production problem, and you wish to acquire new materials to increase production. You then assign a value to each potential new material, according to which would benefit you the most. Which of the following should be reasonable things to expect from these values?

A. The value of each material is nonnegative.

B. The total value of the materials should be as big as possible, to maximize costs associated with their value.

C. The total value of the materials should be as small as possible, to minimize costs associated with their value.

D. The total value of these materials should reflect the value of selling products made with those materials.

E. If a material is not to be used, it has zero value.

Activity 4.1.3 From Exploration 4.1.1, letting x_1 denote the number of healing elixirs, and x_2 denote the amount of poison created. Then, we get that the feasible region satisfies the inequalities:

$$3x_1 + x_2 \leq 34$$
$$x_1 + x_2 \leq 14$$
$$-x_1 \leq 0$$
$$-x_2 \leq 0$$

---

[1]The name Agnesi was chosen by my Spring 2024 class who knew her from her eponymous curve.

which is bounded by hyperplanes:

$$3x_1 + x_2 = 34$$
$$x_1 + x_2 = 14$$
$$-x_1 = 0$$
$$-x_2 = 0$$

with normal vectors

$$\begin{bmatrix} 3 \\ 1 \end{bmatrix}, \begin{bmatrix} 1 \\ 1 \end{bmatrix}, \begin{bmatrix} -1 \\ 0 \end{bmatrix}, \begin{bmatrix} 0 \\ -1 \end{bmatrix}$$

which may be depicted:

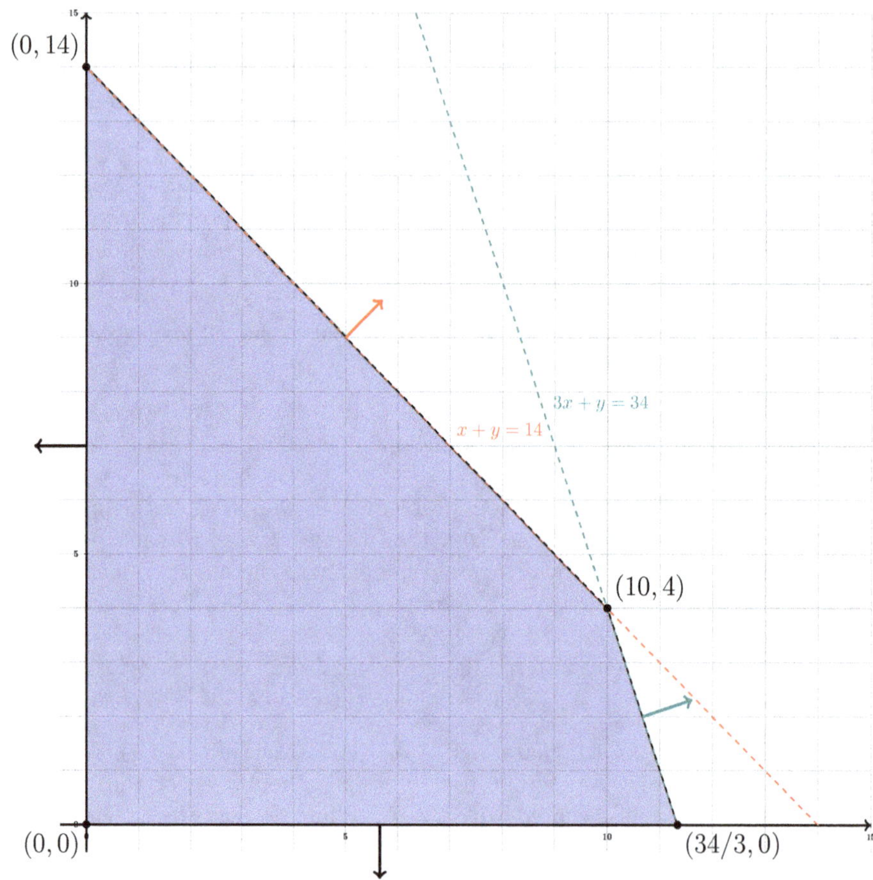

(a) Starting at the basic solution, perform pivots as follows:

x_1	x_2	-1	
3*	1	34	$= -t_1$
1	1	14	$= -t_2$
3	2	0	$= f$

t_1	x_2	-1	
?	?	?	$= -x_1$
?	?*	?	$= -t_2$
?	?	?	$= f$

	t_1	t_2	-1	
	?*	?	?	$= -x_1$
	?	?	?	$= -x_2$
	?	?	?	$= f$

	x_1	t_2	-1	
	?	?	?	$= -t_1$
	?	?	?	$= -x_2$
	?	?	?	$= f$

For each tableau, confirm the solution is feasible.

(b) For each tableau above, if we decrease each decision variable from 0 to -1, how does that change the value of the objective function? What does decreasing a decision variable from 0 to -1 mean geometrically? What does it mean in terms of the normal vectors of the associated intersecting hyperplanes?

(c) Consider that $(0, 14)$ is on the intersection of $-x = 0, x + y = 14$ which are hyperplanes with normal vectors $\begin{bmatrix} -1 \\ 0 \end{bmatrix}$ and $\begin{bmatrix} 1 \\ 1 \end{bmatrix}$ respectively. Note that $3x + 2y = 28$ passes through $(0, 14)$ with normal vector $\begin{bmatrix} 3 \\ 2 \end{bmatrix}$.

Drag sliders for y_1, y_2 so that

$$\begin{bmatrix} 3 \\ 2 \end{bmatrix} = y_1 \begin{bmatrix} -1 \\ 0 \end{bmatrix} + y_2 \begin{bmatrix} 1 \\ 1 \end{bmatrix}.$$

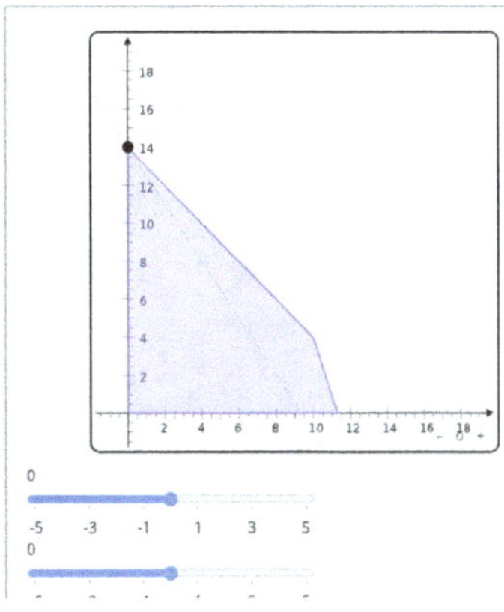

Standalone
Embed

As you adjust y_1, y_2, what do you notice about the resulting linear combination?

(d) For each extreme point, express $\begin{bmatrix} 3 \\ 2 \end{bmatrix}$ as a linear combination of the normal vectors of the corresponding hyperplanes. Then, for each tableau computed above, look at their basic solutions. What point corresponds to each basic solution, and how are these coefficients reflected in each tableau?

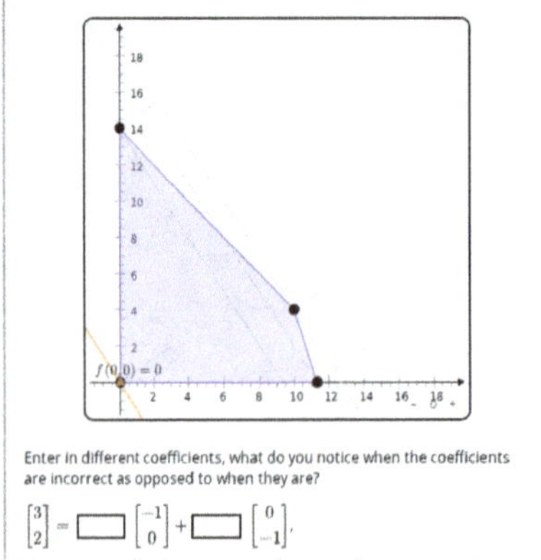

Enter in different coefficients, what do you notice when the coefficients are incorrect as opposed to when they are?

$$\begin{bmatrix} 3 \\ 2 \end{bmatrix} = \boxed{} \begin{bmatrix} -1 \\ 0 \end{bmatrix} + \boxed{} \begin{bmatrix} 0 \\ -1 \end{bmatrix}.$$

(e) For each extreme point in the feasible region, consider the bounding planes who intersect at that point. If you traverse in the direction of the normal vectors from the extreme point, does the objective function increase or decrease? How does this connect to the coefficients we just found?

(f) For which extreme points are the normal vector of the objective plane a linear combination of the normal vectors of intersecting hyperplanes using only positive coefficients? Is there anything special about those extreme points? Is there anything noteworthy about the corresponding tableau?

Activity 4.1.4 Consider the tableau:

x_1	x_2	$\cdots$	x_m	-1	
a_{11}	a_{12}	$\cdots$	a_{1m}	b_1	$= -x_{m+1}$
a_{21}	a_{22}	$\cdots$	a_{2m}	b_2	$= -x_{m+2}$
$\vdots$	$\vdots$	$\ddots$	$\vdots$	$\vdots$	$\vdots$
a_{n1}	a_{n2}	$\cdots$	a_{nm}	b_n	$= -x_{m+n}$
c_1	c_2	$\cdots$	c_m	d	$== f$

Suppose that for $j \in \{1, 2, \ldots, m\}$ each plane $-x_j = 0$ has corresponding normal vector $\mathbf{v}_j$.

(a) Prove that the normal vector for f is $\displaystyle\sum_{j=1}^{m}(-c_j)\mathbf{v}_j$.

(b) Is there anything special about a tableau where f is a nonnegative linear combination of normal vectors?

(c) Suppose that this tableau corresponds to an optimal solution. If we decrease any x_j from 0 to -1, how does f change? What does this decrease correspond to geometrically?

Definition 4.1.5 Recall the canonical maximization problem:

$$\textbf{Maximize: } f(\mathbf{x}) = c_1 x_1 + c_2 x_2 + \cdots c_m x_m - d = \left(\sum_{j=1}^{m} c_j x_j\right) - d$$

$$\textbf{subject to: } a_{11}x_1 + a_{12}x_2 + \cdots a_{1m}x_m \leq b_1$$

$$a_{21}x_1 + a_{22}x_2 + \cdots a_{2m}x_m \leq b_2$$

$$\vdots$$

$$a_{n1}x_1 + a_{n2}x_2 + \cdots a_{nm}x_m \leq b_n$$

$$x_1, x_2, \ldots, x_m \geq 0.$$

The **dual minimization problem** is articulated as follows:

$$\textbf{Minimize: } g(\mathbf{y}) = y_1 b_1 + y_2 b_2 + \cdots y_n b_n - d = \left(\sum_{i=1}^{n} y_i b_i\right) - d$$

$$\textbf{subject to: } a_{11}y_1 + a_{21}y_2 + \cdots a_{n1}y_n \geq c_1$$

$$a_{12}y_1 + a_{22}y_2 + \cdots a_{n2}y_n \geq c_2$$

$$\vdots$$

$$a_{1m}y_1 + a_{2m}y_2 + \cdots a_{nm}y_n \geq c_m$$

$$y_1, y_2, \ldots, y_n \geq 0.$$

We refer to the y_i as **dual variables**. $\diamond$

Activity 4.1.6 Recall Agnesi's business Exploration 4.1.1, and the coefficient computations done in Activity 4.1.3.

(a) Following Definition 4.1.5, write out the dual problem to the maximization problem described in Exploration 4.1.1.

(b) Which of the following best represent the dual variables y_1, y_2 in this context?

    A. The quantity of newt eyes and frogs.

    B. The value of newt eyes and frogs.

    C. The quantity of healing elixirs and poisons.

    D. The value of healing elixirs and poisons.

(c) For each inequality in our dual problem, articulate what those inequalities represent in this context.

(d) Describe the dual objective function in this context.

Activity 4.1.7 Describe three primal maximization problems with some "real world" context, these do not have to be "realistic", they can be fantastical like Agnesi's problem here. Then, describe the dual problem to each and explain what the dual variables mean in each case.

4.2 Duality Theory

In this section, we establish the theoretical underpinnings of duality. This is a proof heavy section.

Observation 4.2.1 Recall the primal maximization problem Activity 2.1.1, and the corresponding dual minimization problem Definition 4.1.5. By letting

$$
A = \begin{bmatrix} a_{11} & a_{12} & \cdots & a_{1m} \\ a_{21} & a_{22} & \cdots & a_{2m} \\ \vdots & \vdots & \ddots & \vdots \\ a_{n1} & a_{n2} & \cdots & a_{nm} \end{bmatrix}, b = \begin{bmatrix} b_1 \\ b_2 \\ \vdots \\ b_n \end{bmatrix}, c^\top = \begin{bmatrix} c_1 & c_2 & \cdots & c_m \end{bmatrix}, d = d.
$$

We can rephrase the primal max problem as follows: Maximize $f = c^\top \mathbf{x} - d$ for $\mathbf{x} \in \mathbb{R}^m$ subject to

$$
A\mathbf{x} \leq b, \mathbf{x} \geq 0.
$$

Here, we understand $\leq, \geq$ to denote entrywise inequality.

Likewise, we can rephrase the dual min problem as follows: Minimize $g = \mathbf{y}^\top b - d$ for $\mathbf{y} \in \mathbb{R}^n$ subject to

$$
\mathbf{y}^\top A \geq c^\top, \mathbf{y} \geq 0.
$$

Activity 4.2.2 In this activity, we explore a foundational relationship between the primal max problem and its dual, called *weak duality*.

(a) Consider the matrix product $\mathbf{y}^\top A\mathbf{x}$. Use this product to show that $g \geq f$.

(b) Suppose there were feasible $\mathbf{x_0}, \mathbf{y_0}$ for which $f(\mathbf{x_0},) = g(\mathbf{y_0})$. What then must be true about these solutions? Can we prove our assertion?

(c) Recall Activity 2.1.12 and Exploration 4.1.1. Consider the primal max and dual min of the associated problems. How does our assertion fit these problems?

(d) Come up with a primal max problem (and corresponding min dual) where $A, b, c, d, \mathbf{x}, \mathbf{y}$ all have *integer* values, so that the primal max and dual min problems achieve optimal solutions $\mathbf{x_0}, \mathbf{y_0}$, where $f(\mathbf{x_0}) < g(\mathbf{y_0})$.

(e) Using the same values for A, b, c, d for the problem we just constructed, suppose we relax the condition that all our values must be integers. What can we say about the optimal solutions then?

What we have proven is the following:

Proposition 4.2.3 Weak Duality. *For a primal maximization problem with objective function f and dual objective g, we have that*

$$
f \leq g.
$$

In particular, if there is a feasible primal solution $\mathbf{x}^*$ and feasible dual solution $\mathbf{y}^*$ such that

$$
f(\mathbf{x}^*) = g(\mathbf{y}^*)
$$

then $\mathbf{x}^*, \mathbf{y}^*$ are optimal solutions for the primal and dual problems respectively.

We have now that if $f = g$ for a pair of feasible solutions, then we have optimality for both problems. It would be good if the converse were also true.

This is encapsulated by the following theorem.

Theorem 4.2.4 The Strong Duality Theorem. *Given a pair of primal max-dual min problems, the primal max problem has an optimal solution* $\mathbf{x}^*$ *if and only if the dual min problem has an optimal solution* $\mathbf{y}^*$. *Moreover,* $f(\mathbf{x}^*) = g(\mathbf{y}^*)$.

To prove this, we recall the idea we explored in Section 4.1 that dual variables were coefficients or weights of normal variables of the bounding hyperplanes.

Activity 4.2.5 Before we delve into the proof, we illustrate the idea with a constrained case. Consider a canonical optimization problem captured by the following tableau (we assume for simplicities sake to let $d = 0$):

x_1	x_2	x_3	x_4	-1	
a_{11}	a_{12}	a_{13}	a_{14}	b_1	$= -t_1$
a_{21}	a_{22}	a_{23}	a_{24}	b_2	$= -t_2$
a_{13}	a_{23}	a_{33}	a_{43}	b_3	$= -t_3$
c_1	c_2	c_3	c_4	0	$= f$

Which after pivoting achieves optimality with the following tableau:

t_3	x_3	x_1	t_2	-1	
a_{11}^*	a_{12}^*	a_{13}^*	a_{14}^*	b_1^*	$= -x_4$
a_{21}^*	a_{22}^*	a_{23}^*	a_{24}^*	b_2^*	$= -t_1$
a_{13}^*	a_{23}^*	a_{33}^*	a_{43}^*	b_3^*	$= -x_2$
c_1^*	c_2^*	c_3^*	c_4^*	d^*	$= f$

(a) Let $y_i^* = -c_?^*$ if $c_?^*$ is the coefficient for t_i in the optimal tableau, and let $y_i^* = 0$ otherwise. Similarly let $s_j^* = -c_?^*$ if $c_?^*$ is the coefficient for x_j in the optimal tableau, and let $s_j^* = 0$ otherwise.

Recall that the optimal tableau represents a reformulation of the original problem where $f = c_1 x_1 + c_2 x_2 + c_3 x_3 + c_4 x_4$ is instead rewritten as $f = c_1^* t_3 + c_2^* x_3 + c_3^* x_1 + c_4^* t_2 - d^*$.

Why must

$$\sum_{j=1}^{4} c_j x_j = \sum_{i=1}^{4} y_i^*(-t_i) + \sum_{j=1}^{4} s_j^*(-x_i) - d^*?$$

(b) Use the fact that $t_i = b_i - \left(\sum_{j=1}^{4} a_{ij} x_j \right)$ to rewrite the above equality without t_i's. Regroup the right hand side so we have the form:

$$\sum_{j=1}^{4} c_j x_j = (\text{linear function in terms of } x_j\text{'s}) + (\text{constant}).$$

(c) Recall that the above equality we established is an equality *as functions of the* x_j's. What must be the constant portion of the right hand side be equal to? What must the linear function in terms of x_j's on the right hand side be equal to?

(d) Recall that $g(\mathbf{y}^*) = \sum_{i=1}^{4} y_i^* b_i$. Prove that $g(\mathbf{y}^*) = -d^*$.

(e) We now need to show that these y_i^*'s we found is a feasible solution to the dual problem. Why do we have that $c_j = \sum_{i=1}^{4} y_i^* a_{ij} - s_j^*$?

(f) Why is each $y_i^*, s_j^* \geq 0$? Why does this show that $\sum_{i=1}^{4} y_i^* a_{ij} \geq c_j$? This shows that s_j's are nonnegative slack variables for the y_i and that the y_i^*'s are a feasible solution.

(g) Show by Weak Duality Proposition 4.2.3 that the y_i^*'s are an optimal dual solution.

Activity 4.2.6 Adopt the arguments of Activity 4.2.5 to prove the Strong Duality Theorem Theorem 4.2.4 for a general primal-dual optimization problem

$$
\begin{array}{cccc|c|c}
x_1 & x_2 & \cdots & x_m & -1 & \\
\hline
a_{11} & a_{12} & \cdots & a_{1m} & b_1 & = -t_1 \\
a_{21} & a_{22} & \cdots & a_{2m} & b_2 & = -t_2 \\
\vdots & \vdots & \ddots & \vdots & \vdots & \vdots \\
a_{n1} & a_{n2} & \cdots & a_{nm} & b_n & = -t_n \\
\hline
c_1 & c_2 & \cdots & c_m & 0 &
\end{array}
$$

which also achieves optimality for some general tableau.

Use the fact that

$$
\sum_{j=1}^{m} c_j x_j = \sum_{i=1}^{n} y_i^*(-t_i) + \sum_{j=1}^{m} s_i^*(-x_i) - d^*,
$$

that $t_i = b_i - \left(\sum_{j=1}^{n} a_{ij} x_j \right)$, and by construction each $y_i^*, s_j^* \geq 0$.

An alternative approach to proving this theorem is provided in Section 9.4.

Activity 4.2.7 Complementary Slackness. We assume the primal problem is canonical.

(a) Prove that $g - f = \mathbf{s}^\top \mathbf{x} + \mathbf{y}^\top \mathbf{t}$.

(b) Suppose $\mathbf{x}$ is a feasible extreme point, and $\mathbf{y}$ and $\mathbf{s}$ are the corresponding coefficients for the normal vectors. If $x_j \neq 0$, what does that say about s_j?

Hint. s_j is the coefficient of the normal vector for the plane $-x_j = 0$. If the feasible solution does not lie on $-x_j = 0$, what can we say about s_j?

(c) The repeat the above for $s_j \neq 0$, $y_i \neq 0$, and $t_i \neq 0$.

(d) Give an example of a linear optimization primal-dual problem, and feasible solutions $\mathbf{x}, \mathbf{y}$ with slack variables $\mathbf{t}, \mathbf{s}$ where $\mathbf{s}^\top \mathbf{x} + \mathbf{y}^\top \mathbf{t} > 0$.

(e) If $\mathbf{x}^*, \mathbf{y}^*$ are feasible optimal solutions with slack variables $\mathbf{t}^*, \mathbf{s}^*$, what must be true about $(\mathbf{s}^*)^\top \mathbf{x}^* + (\mathbf{y}^*)^\top \mathbf{t}^*$?

Definition 4.2.8 Feasible variables $\mathbf{x}, \mathbf{y}$ with slack variables $\mathbf{s}, \mathbf{t}$ are said to exhibit **complementary slackness** if $\mathbf{s}^\top \mathbf{x} = 0, \mathbf{y}^\top \mathbf{t} = 0$. $\diamond$

4.3 Tucker Tableau's, Pivots and Duality

In this section, we examine pivoting with primal-dual tableaus. We will compare (in a good way!) to what we did in Section 2.1.

Activity 4.3.1 Noting that the dual variables y_i are nonnegative weights attached to the hyperplanes defined by $t_i = 0$, and the slack variables for the dual problem s_j are the weights associated with the planes $-x_j = 0$, we can encode all this information in the *Primal-Dual Tucker tableau:*

$$
\begin{array}{c|ccc|c|c}
 & x_1 & \cdots & x_m & -1 & \\
\hline
y_1 & a_{11} & \cdots & a_{1m} & b_1 & = -t_1 \\
\vdots & \vdots & \ddots & \vdots & \vdots & \vdots \\
y_n & a_{n1} & \cdots & a_{nm} & b_n & = -t_n \\
\hline
-1 & c_1 & \cdots & c_m & d & = f \\
 & = s_1 & \cdots & = s_m & = g &
\end{array}
$$

(The additional dividing lines in the top left and bottom right separate the primal decision-slack variables from the dual decision-slack variables.)

(a) Write out the sufficient conditions for the tableau to determine:

    A. The primal problem is feasible.

    B. The dual problem is feasible.

    C. The feasible primal problem is unbounded above.

    D. The feasible dual problem is unbounded below.

    E. The primal problem is infeasible.

    F. The dual problem is infeasible.

    G. The primal problem has a feasible optimal solution.

    H. The dual problem has a feasible optimal solution.

(b) Are any of these conditions identical?

Activity 4.3.2 Recall Activity 1.2.1.

(a) Record this information in a primal-dual Tucker tableau.

(b) Apply the Simplex Algorithm Definition 2.2.11 to this tableau.

(c) Consider the dual solution. What does that mean in the context of the time spent by the painter and the sculptor?

Activity 4.3.3

(a) If a primal problem is infeasible, what *could* be true of the dual problem?

    A. The dual problem has an optimal solution.

    B. The dual problem is unbounded below.

    C. The dual problem is infeasible.

(b) For each of the possibilities discussed in (a), fill in the tableau below to achieve this criteria or explain why it is not possible.

	x_1	x_2	-1	
y_1	-1	?	?	$= -t_1$
y_2	1	?	?	$= -t_2$
?	?	?	?	$= f$
	$= s_1$	$= s_2$	$= g$	

Activity 4.3.4 We now consider unconstrained and equality constrained primal-dual problems.

(a) Suppose for a pair of primal-dual solutions if x_j were allowed to be be any value including negative values, what must be true about s_j? (If x_j is unconstrained, what can we say about the hyperplane $-x_j = 0$ as a bounding hyperplane? Would there ever be an extreme point solution lying on this plane? See interactive below.)

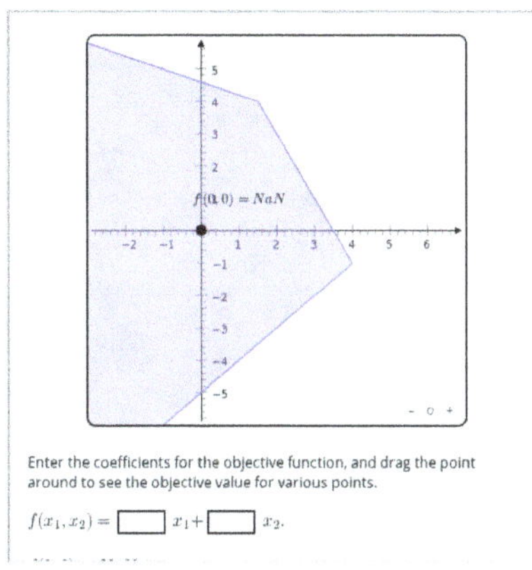

Standalone
Embed

A. s_j could be any value as well. D. $s_j = 0$.

B. $s_j \geq 0$.

C. $s_j \leq 0$. E. $s_j \neq 0$.

(b) Suppose for a pair of primal-dual solutions if $t_i = 0$. What must be true about y_i? (Does it matter whether or not pivoting away from the equality constraint increases or decreases the objective function? See interactive below.)

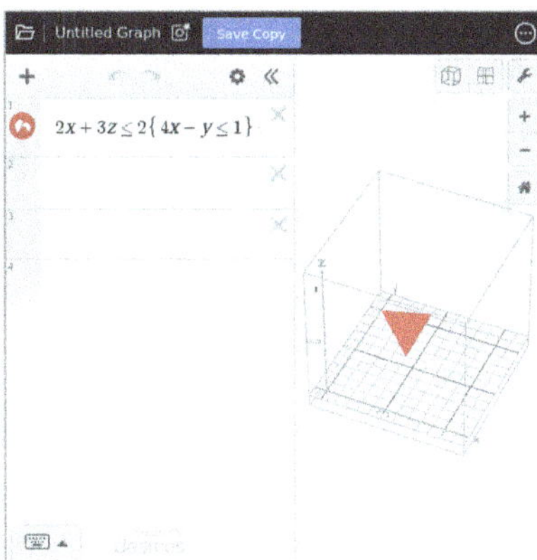

Standalone

    A. y_i could be any value as well.   D. $y_i = 0$.

    B. $y_i \geq 0$.

    C. $y_i \leq 0$.   E. $y_i \neq 0$.

Activity 4.3.5

(a) Solve the following noncanonical primal-dual problem:

	$\boxed{x_1}$	x_2	x_3	-1	
$\boxed{y_1}$	0	-1	-1	-1	$= -0$
y_2	-1	-3	4	0	$= -t_2$
y_3	-1	2	-3	0	$= -t_2$
-1	-1	0	0	0	$= f$
	$= 0$	$= s_2$	$= s_3$	$= g$	

(b) Enter the primal-problem and use Sage to confirm the solution:

```
%display typeset
A = (FIXME)
b = (FIXME)
c = (FIXME)
P = InteractiveLPProblem(A, b, c,
  [FIXME],
  constraint_type = [FIXME],
  variable_type = [FIXME],
  problem_type = FIXME)
P
```

```
print(P.optimal_solution())
print(P.optimal_value())
```

(c) Use Sage to find the dual and solve it:

```
%display typeset
D = P.dual()
D
```

```
print(D.optimal_solution())
print(D.optimal_value())
```

4.4 Summary of Chapter 4

We recall that at each corner or extreme point solution, this solution lies on the intersection of n bounding hyperplanes of the feasible region. We also note that the orthogonal vector for the objective plane may be written as a linear combination of the orthogonal vectors for these bounding hyperplanes, and in particular, the coefficients of the linear combination are exactly the negative of the $c_j's$ in the corresponding tableau.

x_1	x_2	$\cdots$	x_m	-1	
a_{11}	a_{12}	$\cdots$	a_{1m}	b_1	$= -t_1$
a_{21}	a_{22}	$\cdots$	a_{2m}	b_2	$= -t_2$
$\vdots$	$\vdots$	$\ddots$	$\vdots$	$\vdots$	$\vdots$
a_{n1}	a_{n2}	$\cdots$	a_{nm}	b_n	$= -t_n$
c_1	c_2	$\cdots$	c_m	d	

In some sense what $-c_j$ tells us is how much f increases if the corresponding x_j decreases by one, i.e. the bound associated with x_j were relaxed by one. This gives us a way of determining how relaxing each of the constraints compare to each other, and their relative value to each other. So in the case of Exploration 4.1.1, we have:

x_1	x_2	-1	
3	1	34	$= -t_1$
1	1	14	$= -t_2$
3	2	0	$= f$

We have that the normal vector for the objective plane, $\begin{bmatrix} 3 \\ 2 \end{bmatrix}$ may be written

as $\begin{bmatrix} 3 \\ 2 \end{bmatrix} = -3 \begin{bmatrix} -1 \\ 0 \end{bmatrix} + (-2) \begin{bmatrix} 0 \\ -1 \end{bmatrix}$, and lowering x_1 by 1 decreases the objective by 3, and lowering x_2 by 1 decreases the objective by 2.

But should we pivot to:

t_1	t_2	-1	
1/2	$-1/2$	10	$= -t_1$
$-1/4$	3/2	4	$= -t_2$
$-1/2$	$-3/2$	-38	$= f$

Then $\begin{bmatrix} 3 \\ 2 \end{bmatrix} = \frac{1}{2} \begin{bmatrix} 3 \\ 1 \end{bmatrix} + \frac{3}{2} \begin{bmatrix} 1 \\ 1 \end{bmatrix}$ and decreasing t_1 by 1 increases the objective by $\frac{1}{2}$ and decreasing t_2 by 1 increases the objective by $\frac{3}{2}$. Note that this is equivalent to stating how much value would be gained should there by one more newt eye or frog respectively.

This inspires us to define the **dual problem** of a primal maximization problem in Definition 4.1.5, where the y_i are the weights or coefficients of the normal vectors for bounding hyperplanes of the feasible region. We may define the s_j to be the slack variable for each dual constraint so that $\left(\sum_{i=1}^{n} a_{ij} y_i \right) + s_j = c_j x$, or equivalently:

$$\begin{bmatrix} c_1 \\ c_2 \\ \vdots \\ c_m \end{bmatrix} = y_1 \begin{bmatrix} a_{11} \\ a_{12} \\ \vdots \\ a_{1m} \end{bmatrix} + y_2 \begin{bmatrix} a_{21} \\ a_{22} \\ \vdots \\ a_{2m} \end{bmatrix} + \cdots + y_n \begin{bmatrix} a_{n1} \\ a_{n2} \\ \vdots \\ a_{nm} \end{bmatrix}$$

$$+ s_1 \begin{bmatrix} 1 \\ 0 \\ \vdots \\ 0 \end{bmatrix} + s_2 \begin{bmatrix} 0 \\ 1 \\ \vdots \\ 0 \end{bmatrix} + \cdots + s_m \begin{bmatrix} 0 \\ 0 \\ \vdots \\ 1 \end{bmatrix}$$

and we have feasibility if and only if each $y_i, s_j \geq 0$.

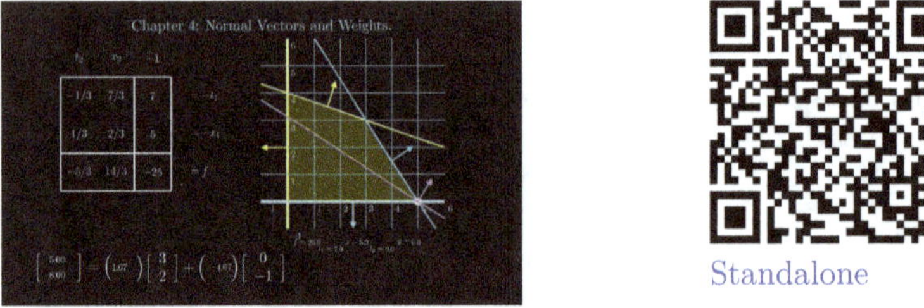

Standalone

Figure 4.4.1 Normal Vectors and Sensitivity Analysis.

We note that by reformulating the primal and dual problem using matrix algebra, then for feasible $\mathbf{x}$ and $\mathbf{y}$, we have that

$$f = \mathbf{c}^\top \mathbf{x} \leq \mathbf{c}^\top A \mathbf{b} \leq \mathbf{y}^\top \mathbf{b} = g$$

which is the statement of the Weak Duality Theorem Proposition 4.2.3. In particular, if $f = g$ for feasible $\mathbf{x}, \mathbf{y}$, then both f, g are optimal.

It turns out the converse to this statement is true as well. To show this, we suppose that f achieves optimality at

$t_?$	$x_?$	$\cdots$	$t_?$	-1	
a^*_{11}	a^*_{12}	$\cdots$	a^*_{1m}	b^*_1	$= -x_?$
a^*_{21}	a^*_{22}	$\cdots$	a^*_{2m}	b^*_2	$= -x_?$
$\vdots$	$\vdots$	$\ddots$	$\vdots$	$\vdots$	$\vdots$
a^*_{n1}	a^*_{n2}	$\cdots$	a^*_{nm}	b^*_n	$= -t_?$
c^*_1	c^*_2	$\cdots$	c^*_m	d^*	

We let $-c_j = t_i, s_j$ depending on which variable c_j is the coefficient for, and by letting each other $y_i, s_j = 0$, we have that

$$f = \sum_{i=1}^{n} c_j x_j - d = \sum_{i=1}^{n} (-y_i) t_i + \sum_{j=1}^{m} (-s_j) x_j - d^*$$

$$= \sum_{i=1}^{n} (-y_i) \left(b_i - \sum_{j=1}^{m} a_{ij} x_j \right) + \sum_{j=1}^{m} (-s_j) x_j - d^*$$

$$= \sum_{j=1}^{m} \left(\left(\sum_{i=1}^{n} y_i a_{ij} \right) - s_j \right) x_j - \left(d^* + \sum_{i=1}^{n} y_i b_i \right)$$

We note that since the above tableau satisfies the conditions for an optimal basic solution, each $y_i, s_j \geq 0$. Since the above equality is an equality of functions of $x_1, \ldots, x_m$, by comparing coefficients of x_j, we have $\sum_{i=1}^{n} y_i a_{ij} \geq c_j$ and $\mathbf{y}$ is a feasible solution. Then we also have that $-d^* = \sum_{i=1}^{n} y_i b_i - d$. The left hand side is the optimal value for f, and the right hand side is the current value of g for this choice of $\mathbf{y}$. So by Proposition 4.2.3, we prove the Strong Duality Theorem Theorem 4.2.4.

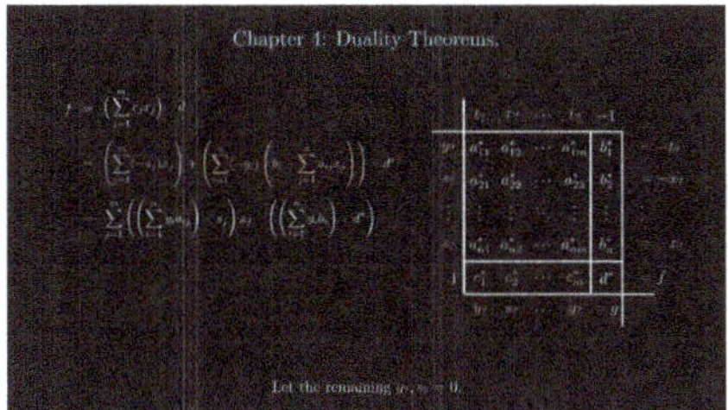

Standalone

Figure 4.4.2 Proofs of the duality theorems.

We then note that the dual problem may be recorded simultaneously as the primal problem within the same tableau

	x_1	$\cdots$	x_m	-1	
y_1	a_{11}	$\cdots$	a_{1m}	b_1	$= -t_1$
$\vdots$	$\vdots$	$\ddots$	$\vdots$	$\vdots$	$\vdots$
y_n	a_{n1}	$\cdots$	a_{nm}	b_n	$= -t_n$
-1	c_1	$\cdots$	c_m	d	$= f$
	$= s_1$	$\cdots$	$= s_m$	$= g$	

and that by using similar arguments as we did in Chapter 2, we may establish conditions for determining the feasibility, optimality, and boundedness if the dual problem. A basic dual variable is feasible if and only if each $c_j \leq 0$, and if we had a $c_j > 0$ but each $a_{ij} \leq 0$ for that problem, then the problem is infeasible, since this inequality can never be satisfied. If the basic dual is feasible then it is also optimal if and only if $b_i \geq 0$. However, if the basic dual is feasible but there is a row i where $a_{ij} \geq 0$, then the dual region is unbounded, and if additionally $b_i < 0$, then the dual problem is unbounded below. These boundedness conditions are sufficient but not necessary, the reader is encouraged to come up with their own counterexamples to necessity.

We conclude by observing that if a primal problem has an unconstrained variable, the corresponding dual variable must be zero, since the associated hyperplane is not a bounding hyperplane. Similarly, if we had an equality constraint in the primal with corresponding slack set to zero, the corresponding dual is allowed to be unconstrained, since it is no longer relevant whether pivoting away from this plane improves the primal objective.

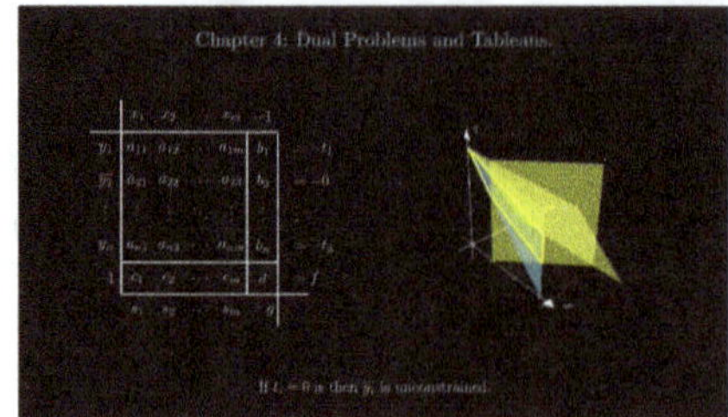

Figure 4.4.3 The dual problem as recorded in tableaus.

4.5 Problems for Chapter 4

Exercises

1. For each problem in Exercise 2.6.5, if the problem has an optimal solution, identify the bounding hyperplanes that the solution lies on, then state the normal vector for the objective plane in terms of the normal vectors of the bounding hyperplanes.

2. Suppose that a company is painting figurines, 4 colors C_1, C_2, C_3, C_4. The colors are mixed from red, green and blue paint.

   C_1 takes 2 oz of red, 2 oz green and 1 oz blue. C_2 takes 3 oz of red, 1 oz green and 1 oz blue. C_3 takes 1 oz of red, 2 oz green and 2 oz blue. C_4 takes 1 oz of red and 4 oz blue.

   Figurines with color C_1, C_3 sell for \$20, C_2 colored figures sell for \$25 dollars and C_4 colored figurines sell for \$30. The company has many figurines to paint and 100 oz each of red, green and blue paint. How many of each colored figurine should be painted to maximize revenue?

   (a) Solve the above canonical maximization problem and restate the objective function in terms of the t_i.

   (b) If we had 110 oz red, 100 oz green, and 100 oz blue paint, without re-solving the problem, what would the optimal revenue be?

   (c) If we had 95 oz red, 100 oz green, and 100 oz blue paint, without re-solving the problem, what would the optimal revenue be?

   (d) If we had 100 oz red, 110 oz green, and 100 oz blue paint, without re-solving the problem, what would the optimal revenue be?

   (e) If we had 100 oz red, 100 oz green, and 95 oz blue paint, without re-solving the problem, what would the optimal revenue be?

3. An office worker is eating in their work cafeteria and they are recently put on a low cholesterol diet. Their usual choices are pasta, fried tofu and chicken sandwiches. The pasta has 6 g of protein, 60 g of carbohydrates, 2 mg of vitamin C, and 60 mg of cholesterol. The fried tofu has 10 g of protein, 40 g of carbohydrates, 2 mg of vitamin C, and 50 mg of cholesterol. The chicken sandwich has 18 g of protein, 40 g of carbohydrates, 2 mg of vitamin C, and 60 mg of cholesterol.

   They need 200 g of protein, 960 g of carbohydrates and 40 mg of vitamin C in a month, and wishes to minimize cholesterol. How many of each meal should they eat?

   (a) Find the optimal solution to the above problem.

   (b) Suppose they needed to increase their protein consumption to 220 g a month. Without recomputing the solution, what would the new minimum cholesterol be?

   (c) Suppose instead that they needed to decrease their carbohydrate consumption to 190 g a month. Without recomputing the solution, what would the new minimum cholesterol be?

   (d) Suppose instead that they needed to increase their vitamin C consumption to 45 g a month. Without recomputing the solution, what would the new minimum cholesterol be?

4. Consider Corollary 9.4.8.

  (a) Prove that if the primal problem is infeasible, and the dual problem is feasible, then the dual problem is unbounded below. (Use Corollary 9.4.8.)

  (b) Prove that if the dual problem is infeasible, and the primal problem is feasible, then the primal problem is unbounded above. (Use **(a)**.)

  (c) Prove that if the primal (dual) problem is feasible and unbounded above (below), then the dual (primal) problem is infeasible.

  (d) Find a primal-dual linear problem where both the primal and dual problems are infeasible.

5. Consider the canonical maximization linear optimization problem below:

$$\textbf{Maximize: } f(x,y) = x + y$$
$$\textbf{subject to: } x + 2y \leq 4$$
$$3x + y \leq 6$$
$$x, y \geq 0.$$

  (a) State the dual canonical minimization linear optimization problem.

  (b) Sketch the constraint sets for both problems above.

  (c) Solve both problems above by applying the simplex algorithm to a dual tableau. Indicate the movement in both constraint set diagrams exhibited by the basic solutions of successive tableaus.

  (d) Is complementary slackness exhibited in the solutions above? Why or why not?

6. For each problem in Exercise 2.6.5, state the dual problem and find its solution.

7. For each of the following, solve the canonical dual minimization problem. If there are infinitely many solutions, classify them all.

(a)

	x_1	x_2	-1	
y_1	1	-2	-2	$= -t_1$
y_2	1	-1	-1	$= -t_2$
-1	1	-2	0	$= f$
	$= s_1$	$= s_2$	$= g$	

(b)

	x_1	x_2	-1	
y_1	-2	1	-1	$= -t_1$
y_2	1	-1	-1	$= -t_2$
-1	2	1	0	$= f$
	$= s_1$	$= s_2$	$= g$	

(c)

	x_1	x_2	-1	
y_1	2	-2	2	$= -t_1$
y_2	-1	1	-1	$= -t_2$
-1	1	1	0	$= f$
	$= s_1$	$= s_2$	$= g$	

(d)

	x_1	x_2	-1	
y_1	6	-1	0	$= -t_1$
y_2	3	1	1	$= -t_2$
y_3	3	1	0	$= -t_3$
-1	2	2	0	$= f$
	$= s_1$	$= s_2$	$= g$	

(e)

	x_1	x_2	x_3	-1	
y_1	1	-1	1	-2	$= -t_1$
y_2	-1	1	1	1	$= -t_2$
-1	0	-1	1	0	$= f$
	$= s_1$	$= s_2$	$= s_3$	$= g$	

(f)

	x_1	x_2	-1	
y_1	-1	0	-1	$= -t_1$
y_2	-1	1	1	$= -t_2$
y_3	-1	-1	-1	$= -t_3$
-1	1	-1	0	$= f$
	$= s_1$	$= s_2$	$= g$	

8. Consider the following tableau:

	x_1	x_2	-1	
y_1	2	0	c_1	$= -t_1$
y_2	0	3	c_2	$= -t_2$
-1	b_1	b_2	0	$= f$
	$= s_1$	$= s_2$	$= g$	

(a) Find b_1, b_2, c_1, c_2 so that the basic solutions of the above tableau are optimal, and both problems have infinitely many solutions.

(b) Classify the choices for b_1, b_2, c_1, c_2 for which the conditions of part **(a)** hold.

9. Consider a two variable, two constraint primal maximization problem.

	x_1	x_2	-1	
y_1	a_{11}	a_{12}	b_1	$= -t_1$
y_2	a_{21}	a_{22}	b_2	$= -t_2$
-1	c_1	c_2	f	$= f$
	$= s_1$	$= s_2$	$= g$	

Recall Weak Duality Proposition 4.2.3, and Complementary Slackness Definition 4.2.8. Suppose we found a pair of primal-dual solutions which satisfied main constraints, but failed one or more nonnegativity constraints.

(a) Must the Weak Duality still hold? Prove or find a counterexample.

(b) Suppose there were a pair of such solutions so that $f = g$, must complementary slackness still hold? Prove or find a counterexample.

10. Prove that for a canonical problem, feasible solutions $\mathbf{x}, \mathbf{y}$, with slack variables $\mathbf{s}, \mathbf{t}$ exhibit complementary slackness if and only if $\mathbf{x}, \mathbf{y}$ are optimal. (This fact is commonly known as the **Complementary Slackness Theorem**.)

11. Solve the following noncanonical primal-dual problems:

(a)

	$\widehat{x_1}$	x_2	-1	
$\widehat{y_1}$	1	1	2	$= 0$
y_2	-1	-1	-2	$= -t_2$
-1	-1	-2	0	$= f$
	$= 0$	$= s_2$	$= g$	

(b)

	$\widehat{x_1}$	x_2	-1	
$\widehat{y_1}$	2	-2	-2	$= 0$
y_2	-1	2	-2	$= -t_2$
-1	-1	-2	0	$= f$
	$= 0$	$= s_2$	$= g$	

(c)

	$\widehat{x_1}$	x_2	-1	
$\widehat{y_1}$	1	-1	-1	$= 0$
y_2	-1	1	-1	$= -t_2$
-1	0	1	0	$= f$
	$= 0$	$= s_2$	$= g$	

		x_1	x_2	x_3	-1	
	y_1	1	-1	1	-1	$= 0$
(d)	y_2	-1	-1	1	1	$= -t_2$
	y_3	-1	1	1	1	$= -t_3$
	-1	1	1	-1	0	$= f$
		$= 0$	$= 0$	$= s_3$	$= g$	

Chapter 5

Zero-Sum Games

We begin the first of several chapters on applications of linear optimization beyond the more straight forward examples we've seen so far, further highlighting the power and versatility of this theory. This particular chapter is about competitive *zero-sum* games. Two players playing against each other in some game with finite choices may find that the efficacy of a choice depends on their opponents choice. A good choice in one scenario may be a poor choice in another. Moreover, if one always makes the same choices, your opponent may be able to anticipate this move, so it makes sense to vary your choices. How should one vary them, and how might your opponent vary theirs? Questions like this are part of an area of mathematics called *Game Theory*.

In Section 5.1 we describe these games and propose a scheme to solve for the optimal strategy for both players. In Section 5.2 we prove that this strategy is valid and that optimal strategies exist. Then in Section 5.3 we see how to apply these principles to games with random components.

We note that in this chapter, we deal quite a bit with randomness and probability. One is encouraged to review basic probability theory in Section A.2 and in particular, the definition of the **expected value** Definition A.2.7.

5.1 Min-Max Games

In this section, we introduce zero-sum games, some basic strategies for approaching them, and highlight a connection to linear optimization.

Exploration 5.1.1 Suppose we have two players, an "Even" player and an "Odd" player. Each player picks an integer from 1-3.

- If the sum is even:

    - If the chosen numbers are distinct, then the Odd player pays the Even player the difference between the numbers.

    - If the chosen numbers are the same, then the Odd player pays the Even player the sum of the numbers.

- If the sum is odd: the Even player pays the Odd player \$3.

(a) Take turns playing this game, do you think either the Even or Odd player has any advantage in this game?

(b) Record the net winnings to the Even player in the following table:

$$Odd$$

		1	2	3
	1	?	?	?
Even	2	?	?	?
	3	?	?	?

(c) Examine this table, and comparing the rows, is there any advantage to the Even player in picking the first row over the third row?

(d) Comparing the columns, is there any advantage to the Odd player in picking the third column over the first column?

(e) Delete any row (column) corresponding to a choice that the Even (Odd) player would never make.

If the Even player always picks a 2, what is the optimal strategy for the Odd player? Similarly if the Odd player always picks a 1, what is the optimal strategy for the even player?

(f) Does either player gain any advantage by picking a single choice and sticking to it?

(g) Suppose the Even player flips a coin to make their choice, if the Odd player picks a 1, what is their average expected winnings? What if they choose a 2?

(h) Suppose the Odd player flips a coin to make their choice, if the Even player picks a 2, what is their average expected winnings? What if they choose a 3?

(i) Does this game favor either the Even or Odd player? What is your best guess?

Definition 5.1.2 In a two-player zero-sum game, where the row player has n choices and the column player has m choices, the **payoff matrix** is a matrix which records in each row and column the net payoff to the row player (this choice is purely by convention, but we will stick to it).

If a row i has entries that are strictly greater than or equal to the entries of another row j, then we say that row i **dominates** row j. We then may delete row j since there is no reason the row player would choose j. Similarly, if column i is less than or equal to column j, column i **dominates** column j and we may delete column j. $\Diamond$

Activity 5.1.3 Consider the payoff matrix for a game between Rowan and Colleen.

$$Colleen$$

	a_{11}	a_{12}	$\cdots$	$\cdots$	a_{1m}
	a_{21}	a_{22}	$\cdots$	$\cdots$	a_{2m}
Rowan	$\vdots$	$\vdots$	$\ddots$	$\ddots$	$\vdots$
	$\vdots$	$\vdots$	$\ddots$	$\ddots$	$\vdots$
	a_{n1}	a_{n2}	$\cdots$	$\cdots$	a_{nm}

(a) Suppose that Rowan pursues a **mixed strategy** a probability distribution of their choices $\mathbf{p}$ where

$$\mathbf{p}^\top = \begin{bmatrix} p_1 \\ p_2 \\ \vdots \\ p_n \end{bmatrix}, p_i \geq 0, \sum_{i=1}^{n} p_i = 1.$$

If Colleen chooses column j as her strategy, what is $E_j(\mathbf{p})$ the expected value of Rowan's earnings?

A. $E_j(\mathbf{p}) = \sum_{i=1}^{n} a_{ij}$. $\qquad$ D. $E_j(\mathbf{p}) = \frac{1}{n} \sum_{i=1}^{n} a_{ij}$.

B. $E_j(\mathbf{p}) = \sum_{i=1}^{n} p_i$. $\qquad$ E. $E_j(\mathbf{p}) = \frac{1}{n} \sum_{i=1}^{n} p_i$.

C. $E_j(\mathbf{p}) = \sum_{i=1}^{n} a_{ij} p_i$. $\qquad$ F. $E_j(\mathbf{p}) = \frac{1}{n} \sum_{i=1}^{n} a_{ij} p_i$.

(b) If Colleen somehow figures out Rowan's strategy $\mathbf{p}$, which column j should she choose?

(c) Rowan is aware that Colleen is a sharp player can likely figure out $\mathbf{p}$. How then should he choose his strategy? $\mathbf{p}$?

A. Maximize $u = \max_{1 \leq j \leq m} E_j(\mathbf{p})$. $\qquad$ C. Minimize $u = \max_{1 \leq j \leq m} E_j(\mathbf{p})$.

B. Maximize $u = \min_{1 \leq j \leq m} E_j(\mathbf{p})$. $\qquad$ D. Minimize $u = \min_{1 \leq j \leq m} E_j(\mathbf{p})$.

(d) At the exact same time, Colleen is pursuing her own mixed strategy $\mathbf{q}$:

$$\mathbf{q} = \begin{bmatrix} q_1 & q_2 & \cdots & q_m \end{bmatrix}, q_j \geq 0, \sum_{j=1}^{m} q_j = 1.$$

If Rowan chooses row i as his strategy, what is $F_i(\mathbf{q})$ the expected value of Colleen's losses?

A. $F_i(\mathbf{q}) = \sum_{j=1}^{m} a_{ij}$. $\qquad$ D. $F_i(\mathbf{q}) = \frac{1}{m} \sum_{j=1}^{m} a_{ij}$.

B. $F_i(\mathbf{q}) = \sum_{j=1}^{m} q_j$. $\qquad$ E. $F_i(\mathbf{q}) = \frac{1}{m} \sum_{j=1}^{m} q_j$.

C. $F_i(\mathbf{q}) = \sum_{j=1}^{m} a_{ij} q_j$. $\qquad$ F. $F_i(\mathbf{q}) = \frac{1}{m} \sum_{j=1}^{m} a_{ij} q_j$.

(e) If Rowan somehow figures out Colleen's strategy $\mathbf{q}$, which row i should he choose?

(f) Colleen is also aware that Rowan is a sharp player can likely figure out **q**. How then should she choose her strategy? **q**?

A. Maximize $v = \max\limits_{1 \leq i \leq n} F_i(\mathbf{q})$.   C. Minimize $v = \max\limits_{1 \leq i \leq n} F_i(\mathbf{q})$.

B. Maximize $v = \min\limits_{1 \leq i \leq n} F_i(\mathbf{q})$.   D. Minimize $v = \min\limits_{1 \leq i \leq n} F_i(\mathbf{q})$.

(g) Consider the primal-dual problems encoded by the tableau:

	$⊚v$	q_1	q_2	$\cdots$	q_m	-1	
$⊚u$	0	-1	-1	$\cdots$	-1	-1	$= -0$
p_1	-1	a_{11}	a_{12}	$\cdots$	a_{1m}	0	$= -t_1$
p_2	-1	a_{21}	a_{22}	$\cdots$	a_{2m}	0	$= -t_2$
$\vdots$	$\vdots$	$\vdots$	$\vdots$	$\ddots$	$\vdots$	$\vdots$	$\vdots$
p_n	-1	a_{n1}	a_{n2}	$\cdots$	a_{nm}	0	$= -t_n$
-1	-1	0	0	$\cdots$	0	0	$= f$
	$= 0$	$= s_1$	$= s_2$	$\cdots$	$= s_n$	$= g$	

Write out both the primal and dual problems encoded by this tableau. (Including all equalities, inequalities, and the objective functions)

(h) What primal constraint does the first row represent? How does it relate to Colleen's strategy?

(i) What primal constraint do the next n rows represent? How does it relate to Colleen's strategy?

(j) What is the primal objective function? How does it relate to Colleen's strategy?

(k) Repeat **(h)**-**(j)** for the columns, and with regards to Rowan's strategy.

(l) Supposing that this system has an optimal primal and dual solution, what would those solutions represent?

Definition 5.1.4 Suppose that the reduced payoff matrix had an entry a_{ij} that is the largest value in its column and the smallest value in its row. Such an entry a_{ij} is called a **saddle point**. $\diamond$

Activity 5.1.5 Suppose a reduced payoff matrix had a saddle point a_{ij}, and consider the resulting tableau:

	$⊚v$	q_k	$\cdots$	q_j	-1	
$⊚u$	0	-1	$\cdots$	-1	-1	$= -0$
p_ℓ	-1	$a_{\ell k}$	$\cdots$	$a_{\ell j}$	0	$= -t_\ell$
$\vdots$	$\vdots$	$\vdots$	$\ddots$	$\vdots$	$\vdots$	$\vdots$
p_i	-1	a_{ik}	$\cdots$	a_{ij}	0	$= -t_i$
-1	-1	0	$\cdots$	0	0	$= f$
	$= 0$	$= s_k$	$\cdots$	$= s_j$	$= g$	

(a) Pivot first on the entry with a $*$ then $**$.

	$\circledv$	q_k	$\cdots$	q_j	-1	
$\circledu$	0	-1	$\cdots$	-1^*	-1	$= -0$
p_ℓ	-1	$a_{\ell k}$	$\cdots$	$a_{\ell j}$	0	$= -t_\ell$
$\vdots$	$\vdots$	$\vdots$	$\ddots$	$\vdots$	$\vdots$	$\vdots$
p_i	-1^{**}	a_{ik}	$\cdots$	a_{ij}	0	$= -t_i$
-1	-1	0	$\cdots$	0	0	$= f$
	$= 0$	$= s_k$	$\cdots$	$= s_j$	$= g$	

	t_i	q_k	$\cdots$	0	-1	
s_j	$?$	$?$	$\cdots$	$?$	C	$= -q_j$
p_ℓ	$?$	$?$	$\cdots$	$?$	C'	$= -t_\ell$
$\vdots$	$\vdots$	$\vdots$	$\ddots$	$\vdots$	$\vdots$	$\vdots$
0	$?$	$?$	$\cdots$	$?$	$?$	$= -\circledv$
-1	B	B'	$\cdots$	$?$	D	$= f$
	$= p_i$	$= s_k$	$\cdots$	$= \circledu$	$= g$	

(b) For each entry B, B', C, C' determine if:

A. The entry is zero.

B. The entry is positive.

C. The entry is negative.

D. The entries value cannot be determined.

(c) What is D? What are $\circledu$ and $\circledv$?

(d) After we delete the appropriate rows and columns, what could be said about the resulting primal-dual problems?

(e) Would it make a difference if we pivoted by $**$ first then $*$?

Activity 5.1.6

(a) Follow the outline provided by Activity 5.1.3 to find the optimal strategies for the Even and Odd players in Exploration 5.1.1, and who if anyone the game favors.

(b) To test out this solution edit this R code:

```r
numsamp = 1000000

Rowan = sample(1:3, numsamp, replace = TRUE, prob =
    c(FIXMER1, FIXMER2, FIXMER3))
Colleen = sample(1:3, numsamp, replace = TRUE, prob =
    c(FIXMEC1, FIXMEC2, FIXMEC3))

PayoffMatrix = matrix(c(2,-3,2,-3,4,-3,2,-3,6), nrow =
    3, ncol = 3)

Winnings = rep(NA, numsamp)

for(i in 1:numsamp){
    Winnings[i]=PayoffMatrix[Rowan[i],Colleen[i]]
}

hist(Winnings)
abline(v=mean(Winnings), col="red",lwd=3, lty=2 )
print(mean(Winnings))
```

Where `FIXMER1`, `FIXMER2`, `FIXMER3` represent the entries for the optimal mixed strategy for the row player Even, and `FIXMEC1`, `FIXMEC2`, `FIXMEC3` are for the optimal mixed strategy for the column player Odd.

Then run the cell and see the distributions of winnings and the average winnings. How does this value compare to what you found?

(c) Taking turns, have one student pick a new strategy for Even, and another student then modify the strategy for Odd in light of the new strategy. Can we do better than Odd's current best strategy?

(d) Conversely, taking turns, one student pick a new strategy for Odd, and another student then modify the strategy for Even in light of the new strategy. Can we do better than Even's current best strategy?

Activity 5.1.7 Find the optimal strategies for two players Rowan and Colleen playing "Rock, Paper, Scissors".

5.2 von Neumann Minimax Theorem

In this section, show that the linear optimization scheme from Section 5.1 gives us exactly what we want by proving von Neumann's Theorem.

Exploration 5.2.1 Consider the tableau in Task 5.1.3.g and the associated primal-dual problems. Which of the following could possibly be true for these problems?

    A. Both primal and dual problem achieve optimality.

    B. The primal problem is unbounded and the dual problem is infeasible.

    C. The primal problem is infeasible and the dual problem is unbounded.

    D. Both problems are infeasible.

It would be very bad if either problem were infeasible or unbounded! It would be good to show that this is not the case.

Activity 5.2.2 Let A be a payoff matrix and $\mathbf{p}, \mathbf{q}$ represent the strategies of the row and column players respectively, with feasible regions $F_{\mathbf{p}}, F_{\mathbf{q}}$.

Let $A_{(i)}$ denote the ith row of a matrix A and let $A^{(j)}$ denote the jth column of A.

(a) Given a fixed column strategy $\mathbf{q}'$ which of these describes the role for the row player?

    A. $\displaystyle\min_{\mathbf{p}\in F_{\mathbf{p}}} \mathbf{p}^{\top} A\mathbf{q}'.$

    B. $\displaystyle\max_{\mathbf{p}\in F_{\mathbf{p}}} \mathbf{p}^{\top} A\mathbf{q}'.$

    C. $\displaystyle\min_{\mathbf{p}\in F_{\mathbf{p}}} \sum_{i=1}^{n} \mathbf{p}_i A_{(i)}\mathbf{q}'.$

    D. $\displaystyle\max_{\mathbf{p}\in F_{\mathbf{p}}} \sum_{i=1}^{n} \mathbf{p}_i A_{(i)}\mathbf{q}'.$

(b) Given a fixed row strategy $\mathbf{p}'$ which of these describes the role for the column player?

    A. $\displaystyle\min_{\mathbf{q}\in F_{\mathbf{q}}} \mathbf{p}'^{\top} A\mathbf{q}.$

    B. $\displaystyle\max_{\mathbf{q}\in F_{\mathbf{q}}} \mathbf{p}'^{\top} A\mathbf{q}.$

    C. $\displaystyle\min_{\mathbf{p}\in F_{\mathbf{p}}} \sum_{j=1}^{m} \mathbf{p}^{\top} A^{(j)}\mathbf{q}_j.$

    D. $\displaystyle\max_{\mathbf{p}\in F_{\mathbf{p}}} \sum_{j=1}^{m} \mathbf{p}^{\top} A^{(j)}\mathbf{q}_j.$

Activity 5.2.3 We prove an interesting way to think of the optimal strategies.

Let $\mathbf{q}'$ denote a fixed column strategy, let $u_1 = \max_{\mathbf{p}\in F_{\mathbf{p}}} \mathbf{p}^{\top} A\mathbf{q}'$, and let $u_2 = \max_{1\le i\le n} A_{(i)}\mathbf{q}'.$

(a) Recall that $\mathbf{p}^{\top} A\mathbf{q}' = \displaystyle\sum_{i=1}^{n} \mathbf{p}_i A_{(i)}\mathbf{q}'$. Prove that $\mathbf{p}^{\top} A\mathbf{q}' \le u_2$.

(b) Why must $u_1 \le u_2$?

(c) Show that there is a (very simple) row strategy $\mathbf{p}''$ where $(\mathbf{p}'')^{\top} A\mathbf{q}' = u_2$.

(d) Why must $u_2 \le u_1$?

(e) What have we proven?

Activity 5.2.4 We now prove a characterization theorem about the optimal solutions for both the row and column player.

Suppose we have a payoff matrix A where every entry is positive. In other words, after each round Rowan is guaranteed to win money and Colleen is guaranteed to lose money. Rowan's strategy here is to take Colleen for as much money as he can and Colleen's strategy is to minimize her losses.

(We'll ignore the obvious question of why Colleen would be willing to play this game.)

(a) Write out the primal maximization problem for the Linear Optimization formulation of this game:

	v	q_1	q_2	$\cdots$	q_m	-1	
u	0	-1	-1	$\cdots$	-1	-1	$= -0$
p_1	-1	a_{11}	a_{12}	$\cdots$	a_{1m}	0	$= -t_1$
p_2	-1	a_{21}	a_{22}	$\cdots$	a_{2m}	0	$= -t_2$
$\vdots$	$\vdots$	$\vdots$	$\vdots$	$\ddots$	$\vdots$	$\vdots$	$\vdots$
p_n	-1	a_{n1}	a_{n2}	$\cdots$	a_{nm}	0	$= -t_n$
-1	-1	0	0	$\cdots$	0	0	$= f$
	$= 0$	$= s_1$	$= s_2$	$\cdots$	$= s_n$	$= g$	

Write out the noncanonical primal problem including the objective function and constraint equalities and inequalities involving the q_i and v where appropriate. (There should be no slack variables here.)

(b) For each row and column strategy, we have expected row winnings u and v respectively. Why are these always positive?

(c) Consider the inequality constraints in our formulation, divide each of these by v. Let $\tilde{q}_j := \frac{q_j}{v}$. Can we rewrite our inequalities as linear combinations of $\tilde{q}_j$ is less than or equal to some constant?

(d) Consider the equality constraint after dividing by v, rewrite this equality in terms of $\tilde{q}_j$ without negatives.

Remember, Colleen's strategy is to minimize v which must be positive. Can we rephrase this as maximizing or minimizing a linear function involving then $\tilde{q}_j$? What is this linear function and is it a maximization or minimization problem? (Note that the solution to this problem likely isn't the solution to the original problem, but both are optimized under the same conditions.)

(e) Rewrite the new *canonical* linear optimization problem with variables $\tilde{q}_j$ that optimizes Colleen's strategy.

(f) Why is the feasible region for Colleen's new problem nonempty but bounded? What does the Extreme Value Theorem then say about this?

(g) Repeat tasks **(b)-(c)** for Rowan's strategy, where $\tilde{p}_i = \frac{p_i}{u}$.

(h) Compare Rowan and Colleen's problems with the $\tilde{p}_i, \tilde{q}_j$. Show that these problems are dual problems to each other. Which is the primal max and which is the dual min?

(i) What does the Strong Duality Theorem Theorem 4.2.4 say about the optimal solutions to both problems? What, in turn, does that say about $\textcircled{u}$ and $\textcircled{v}$?

(j) We're still in this pretty ridiculous situation where Colleen is for some reason willing to throw money away at Rowan. To balance things out, Rowan has to pay Colleen \$5 after each round. Would this fact change anything about Rowan and Colleen's strategies? Say Rowan paid Colleen \$r, would that make any difference?

(k) Let $\mathbf{p}, \mathbf{q}$ denote any strategy for Rowan and Colleen. Let E denote a $n \times m$ matrix with all 1's. Show that $\mathbf{p}^\top (rE)\mathbf{q} = r$.

(l) Show that for fixed strategies $\mathbf{p}', \mathbf{q}'$ and not fixed strategies $\mathbf{p}, \mathbf{q}$ that $\mathbf{p}^\top (A + rE)\mathbf{q}'$ is maximized when $\mathbf{p}^\top A\mathbf{q}'$ is maximized and $\mathbf{p}'^\top (A + rE)\mathbf{q}$ is minimized when $\mathbf{p}'^\top A\mathbf{q}$ is minimized.

Theorem 5.2.5 von Neumann's Minimax Theorem. *Let A be a payoff matrix and $\mathbf{p}, \mathbf{q}$ represent the strategies of the row and column players respectively, with feasible regions $F_\mathbf{p}$ and $F_\mathbf{q}$. Also let $A_{(i)}$ denote the ith row of a matrix A and let $A^{(j)}$ denote the jth column of A.*

Then, there are optimal strategies $\mathbf{p}', \mathbf{q}'$ such that:

$$\min_{\mathbf{q} \in F_\mathbf{q}} \mathbf{p}'^\top A\mathbf{q} = \min_{1 \le i \le n} A_{(i)}\mathbf{q} = \mathbf{p}'^\top A\mathbf{q}' = \max_{1 \le j \le m} \mathbf{p}^\top A^{(j)} = \max_{\mathbf{p} \in F_\mathbf{p}} \mathbf{p}^\top A\mathbf{q}'.$$

*We call the described value the **von Neumann value** of the game.*

Activity 5.2.6 Revisit Exploration 5.2.1. Has your answer adjusted at all?

Activity 5.2.7 Consider the payoff matrix $A = \begin{bmatrix} 1 & 0 \\ 1 & 2 \end{bmatrix}$.

(a) Find the optimal strategy $\mathbf{q}'$ for Colleen in this game, and the game value v.

(b) Find a strategy $\mathbf{p}$ for Rowan so that $\mathbf{p}^T A\mathbf{q}' = v$, but $\mathbf{p}$ is not the optimal strategy.

(c) What does this say about Theorem 5.2.5?

Activity 5.2.8 In a simplified game of battleship played on a 2×3 board, Colleen selects two consecutive squares on the board to place her ship. Rowan then picks one of six squares to fire at. If he hits, he gets a point, otherwise Colleen gets a point.

(a) Write out a payoff matrix for this game. (Why is it 6×7?)

(b) Find the optimal solution for Colleen using Sage:

```
%display typeset
A = (FIX_ME)
b = (FIX_ME)
c = (FIX_ME)
P = InteractiveLPProblem(A, b, c,
 ["v","q1","q2","q3","q4","q5","q6","q7"],
  constraint_type = ["==", "<=", "<=", "<=", "<=", "<=",
      "<="],
  variable_type = ["", ">=", ">=", ">=", ">=", ">=",
      ">="])
P
```

```
print(P.optimal_solution())
print(P.optimal_value())
```

Does the game favor Rowan or Colleen?

(c) Use Sage to find the optimal solution for Rowan:

```
%display typeset
D = P.dual()
D
```

```
print(D.optimal_solution())
print(D.optimal_value())
```

5.3 Games of Chance

For games with random components, such as gambling, we can still apply our techniques, using the expected winnings. This requires some work, and we explore how to do so.

Activity 5.3.1 Consider the following game. Rowan and Colleen place $25 into a pot. Then they are dealt either a Jack, Queen or King at random. This deck only has those three cards. Whoever has the highest card takes the pot.

Rowan has two options. He may raise by $10 or fold. If he folds, he loses $25. Otherwise Colleen then can either fold or call. If she folds, she loses $25, if she calls, she puts $10 into the pot.

Note that if Rowan folds, even if Colleen had planned on folding, she would win the $25.

(a) What would best describe possible choices of strategy for Rowan and Colleen?

    A. The cards Jack, Queen, King.

    B. Whether to fold/raise for Rowan, whether to fold/call for Colleen.

    C. Whether to fold/raise for Rowan depending on the card he is dealt, whether to fold/call for Colleen, depending on the card she is dealt.

(b) List the possible pairs of Rowan/Colleen hands.

(c) Let's say Rowan raises on a Jack, folds on a Queen, and raises on a King, denoted as RFR. Let's saw Colleen folds on a Jack, and calls on a Queen or King, denoted FCC.

If both players are committed to these strategies, what are Rowan's expected net winnings? (Note that all the above hand pairs are equally likely, what are Rowan's net winnings in each case?)

(d) If Rowan's strategy is (for some reason) FFF, what are Rowan's net winnings?

(e) If Colleen's strategy is FFF, what are Rowan's net winnings? (These may be different for each of Rowan's choice of strategy.)

(f) Without computing the entire payoff matrix, are there any obviously poor strategies for Rowan or Colleen?

(g) Fill out the remainder of this payoff matrix, where the entries are expected values.

	FFF	FFC	FCF	CFF	FCC	CFC	CCF	CCC
FFF	?	?	?	?	?	?	?	?
FFR	?	$-25/3$	$-20/3$	$-20/3$	$-20/3$	$-20/3$	-5	-5
FRF	?	$-55/3$	$-25/3$	$-20/3$	$-55/3$	$-50/3$	$-20/3$	$-50/3$
RFF	?	$-55/3$	$-55/3$	$-25/3$	$-85/3$	$-55/3$	$-55/3$	$-85/3$
FRR	?	$-5/3$	10	$35/3$	0	$5/3$	$40/3$	$10/3$
RFR	?	$-5/3$	0	10	?	0	$5/3$	$-25/3$
RRF	?	$-35/3$	$-5/3$	?	$-65/3$	-10	0	-20
RRR	?	5	$50/3$	$85/3$	$-10/3$	$25/3$	20	?

(h) After dominating what does this table reduce to?

	???	???
???	?	?
???	?	?

(i) Solve for the optimal strategies using the method of your choice.

(j) Who does the game favor and by how much?

(k) If Rowan is dealt a Jack, what is his optimal strategy (as a pair of probabilities to raise or fold). Queen? King?

(l) If Colleen is dealt a Jack, what is her optimal strategy. Queen? King?

Activity 5.3.2 We introduce a second game here. Each player places $b into the pot. Then the each secretly flip a coin. We consider heads "greater" than tails.

Rowan then has the options of pass or bet. If he passes, then both players reveal their coin and the higher value wins. If they are the same, players split the pot evenly, and both players net wins/losses are $0. His other option is to bet, in which case he adds $r to the pot.

Then Colleen has a choice as well, to fold or call. If she folds, then Rowan nets the $b. Otherwise, she calls, and also adds $r to the pot and both coins are revealed.

(a) Suppose that Rowan will stick to the strategy of RP (raise on head, pass on tails) and Colleen choses CC (call on both heads or tails). What are Rowan's expected winnings in this case?

(b) Fill out the payoff matrix for this game.

	FF	FC	CF	CC
PP	?	0	0	?
PR	?	$b/2$	$(b-r)/4$	?
RP	$b/4$	?	0	?
RR	b	?	?	0

(c) It's not possible to determine all the domination without knowing what b, r are. However, knowing $b, r > 0$, dominate as much as possible

	??	??
??	?	?
??	?	?

(d) If $b \leq r$, use domination to find the optimal pure strategy for both players.

(e) If $b > r$, use linear optimization methods to find the optimal mixed strategies for both players.

5.4 Summary of Chapter 5

In this chapter, we begin by introducing the notion of zero-sum games. Suppose two players, Rowan and Colleen had n and m strategies to choose from, and given an i, j choice of strategies, the net payout to Rowan was a_{ij} (with a negative value meaning a payout to Colleen). This may be recorded in what is called a **payoff matrix**.

$$
\begin{array}{c}
\text{Colleen} \\
\text{Rowan} \quad
\begin{bmatrix}
a_{11} & a_{12} & \cdots & \cdots & a_{1m} \\
a_{21} & a_{22} & \cdots & \cdots & a_{2m} \\
\vdots & \vdots & \ddots & \ddots & \vdots \\
\vdots & \vdots & \ddots & \ddots & \vdots \\
a_{n1} & a_{n2} & \cdots & \cdots & a_{nm}
\end{bmatrix}
\end{array}
$$

We note that some strategies may be simply bad choices for either player. For example, if there were two rows i, i' where $a_{ij} \geq a_{i'j}$, for each j, then there is no reason for Rowan to pick i' over i and we may delete the i' row. Similarly if for columns j, j', $a_{ij} \leq a_{ij'}$, we may delete the j' column. This process of deleting rows and columns is called **domination**.

Once domination is complete, we can find the resulting optimal strategies by considering the primal-dual optimization problem:

	$\textcircled{v}$	q_1	q_2	$\cdots$	q_m	-1	
$\textcircled{u}$	0	-1	-1	$\cdots$	-1	-1	$= -0$
p_1	-1	a_{11}	a_{12}	$\cdots$	a_{1m}	0	$= -t_1$
p_2	-1	a_{21}	a_{22}	$\cdots$	a_{2m}	0	$= -t_2$
$\vdots$	$\vdots$	$\vdots$	$\vdots$	$\ddots$	$\vdots$	$\vdots$	$\vdots$
p_n	-1	a_{n1}	a_{n2}	$\cdots$	a_{nm}	0	$= -t_n$
-1	-1	0	0	$\cdots$	0	0	$= f$
	$= 0$	$= s_1$	$= s_2$	$\cdots$	$= s_n$	$= g$	

Where the p_i, q_j represent probabilities for Rowan and Colleen and $\textcircled{u}, \textcircled{v}$ represent the values of Rowan and Colleens strategies respectively.

We note that a **saddle point**, is an entry a_{ij} that is the smallest value in row i but the greatest in column j. If there is a saddle point, the optimal strategies for Rowan and Colleen are i and j respectively.

Otherwise, if there are no saddle points, then both players need to employ a random mix of strategies, and solving the above primal-dual problem find $\mathbf{p}, \mathbf{q}$, the probability distribution of the valid strategies for both players.

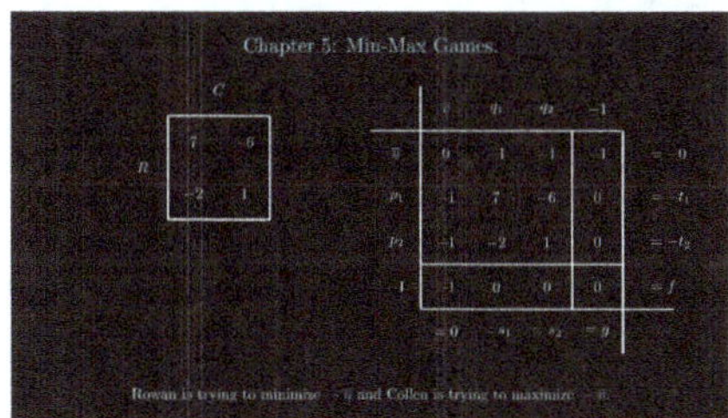

Figure 5.4.1 Zero-Sum Games.

The theorem that shows this approach is valid is the **von Neumann minimax Theorem** Theorem 5.2.5. We first note that technically the above problem attempts to maximize u (minimize v) constrained by $u \leq \min_{1 \leq i \leq n} A_i \mathbf{q}$, (minimize $v \geq \min_{1 \leq j \leq m} \mathbf{p} A^j$), i.e. it maximizes or minimizes across pure strategies, but we wish to max or min across mixed strategies. These can be shown to be equivalent, a pure strategy is a type of mixed strategy, and the other inequality can be shown through some algebra.

Then if we assume that each $a_{ij} > 0$, then by letting $\tilde{p}_i = \frac{p_i}{u}, \tilde{q}_j = \frac{q_j}{v}$, the above system is equivalent to solving:

	$\tilde{q}_1$	$\tilde{q}_2$	$\cdots$	$\tilde{q}_m$	-1	
$\tilde{p}_1$	a_{11}	a_{12}	$\cdots$	a_{1m}	1	$= -t_1$
$\tilde{p}_2$	a_{21}	a_{22}	$\cdots$	a_{2m}	1	$= -t_2$
$\vdots$	$\vdots$	$\vdots$	$\ddots$	$\vdots$	$\vdots$	$\vdots$
$\tilde{p}_n$	a_{n1}	a_{n2}	$\cdots$	a_{nm}	1	$= -t_n$
-1	1	1	$\cdots$	1	0	$= f$
	$= s_1$	$= s_2$	$\cdots$	$= s_n$	$= g$	

We note that since each $a_{ij} > 0$, then the primal region is bounded, and so by the Extreme Value Theorem, the primal problem achieves a maximum, and so by the Strong Duality Theorem, the dual achieves optimality as well.

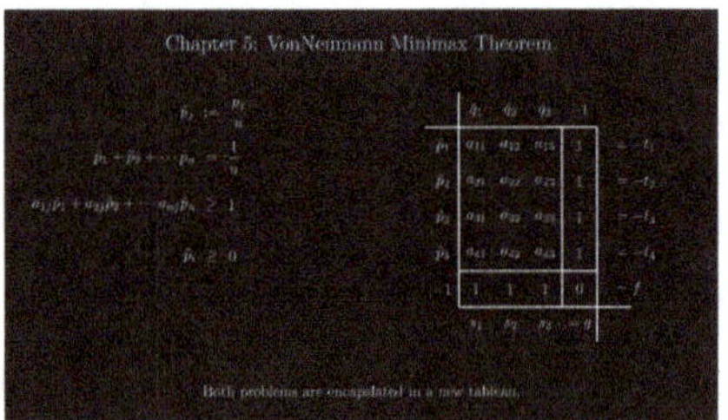

Figure 5.4.2 Proof of the von Neumann minimax Theorem.

Finally, we note that when playing games of chance, the random element prevents us from knowing the actual payouts of different strategies. But we can decide to maximize/minimize expected payouts, and then may proceed as before.

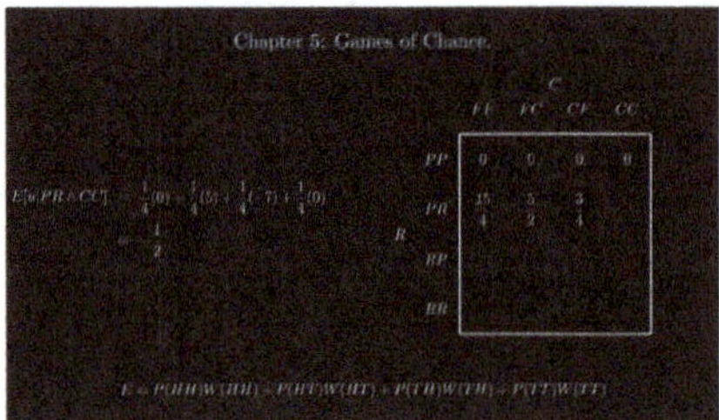

Figure 5.4.3 Games of Chance.

5.5 Problems for Chapter 5

Exercises

1. For the following games, write out the payoff matrix and the reduced matrix after domination.

   (a) Rowan has a 3 of hearts, 4 of spades, 9 of spades and 10 of hearts. Colleen has a 5 of spades, 6 of hearts, 7 of hearts and 8 of spades.

   Each player selects a card without revealing it, and both players flip their cards over at the same time. If the suits are the same, then Rowan wins the sum of the two card values. Otherwise, Colleen wins that sum.

   (b) The same as **(a)** but reverse the payoff conditions for Rowan and Colleen.

   (c) Rowan picks an even integer from 1-6, and Colleen picks an odd integer from 1-6. If the difference is less than 3, the player who played the bigger number wins the sum of the two values. Otherwise the player who played the smaller number wins the sum of the two values.

   (d) Rowan picks an integer from 1-3. Colleen picks two guesses which may be the same. Colleen reveals her guesses one at a time. If she guesses correctly, she wins a number of dollars equal to the value of their other guess. If she does not guess correctly, Rowan wins a number of dollars equal to the sum of both guesses.

   (e) *Colonel Blotto.*

   Colonel Rowan is attacking a town defended by Colonel Colleen. Rowan has three regiments and Colleen has four. There are two routes to the town. Rowan and select any number of regiments to attack along each route, up to a total of three and Colleen can likewise assign her four regiments along either route, neither knowing beforehand the other's strategy.

   Along each route, whatever side has a greater number of regiments wins points equal to the number of regiments sent by the opposing side (as they capture those units). Furthermore, if Rowan wins along either route, he captures the town also worth one point.

   (f) *Morra.*

   Rowan and Colleen each simultaneously hold up one or two fingers and shouts a guess for the total number of fingers held. If either Rowan or Colleen guess correctly, then they collect dollars from their opponent equal to this number of fingers. If they both guess correctly or both guess incorrectly, then no money changes hands.

2. For each of the following payoff matrices, determine if there is a value of x so that the matrix has a saddle point. If so, determine the value(s) of x and the saddle point(s).

(a) $\begin{bmatrix} 1 & 2 \\ x & 3 \end{bmatrix}$

(b) $\begin{bmatrix} 1 & 2 \\ 3 & x \end{bmatrix}$

(c) $\begin{bmatrix} 1 & 3 \\ 2 & x \end{bmatrix}$

(d) $\begin{bmatrix} 3 & x \\ x & 1 \end{bmatrix}$

(e) $\begin{bmatrix} 2 & 1 \\ 3 & x \end{bmatrix}$

(f) $\begin{bmatrix} 2 & 3 \\ 1 & x \end{bmatrix}$

(g) $\begin{bmatrix} x & 1 \\ 2 & 3 \end{bmatrix}$

(h) $\begin{bmatrix} x & 2 \\ 1 & 3 \end{bmatrix}$

3. Write out a reduced payoff matrix with exactly three saddle points.

4. For each of the following payoff matrices, find the von Neumann value and optimal strategy for the payoff matrix.

(a) $\begin{bmatrix} 0 & 1 & 0 & -2 & -1 \\ 3 & 1 & 1 & -2 & 2 \\ 1 & 1 & 0 & -2 & 0 \\ 5 & -3 & -2 & 4 & 0 \\ 2 & 1 & 0 & -3 & 0 \end{bmatrix}$

(b) $\begin{bmatrix} 2 & 2 & -1 & -1 & 2 & 2 \\ -7 & -6 & 5 & 5 & -6 & -2 \\ 1 & 1 & 0 & -1 & 2 & 2 \\ 2 & 2 & 0 & -1 & 2 & 2 \\ 6 & -7 & 5 & 4 & -6 & -2 \end{bmatrix}$

(c) $\begin{bmatrix} 4 & -5 & -7 & 4 \\ -1 & 4 & 4 & 0 \\ -2 & 3 & 4 & 0 \\ -2 & 4 & 4 & -1 \end{bmatrix}$

$$
\text{(d)} \quad \begin{bmatrix} 1 & 2 & -1 & 1 & 1 & 0 \\ 3 & 2 & 0 & 1 & 3 & 1 \\ 2 & 2 & 0 & 0 & 1 & 0 \\ 2 & 1 & 0 & 1 & 1 & 0 \\ 2 & 1 & 0 & -1 & 2 & 1 \\ 1 & 1 & 0 & 1 & 1 & 2 \end{bmatrix}
$$

$$
\text{(e)} \quad \begin{bmatrix} -5 & 1 & 3 & -1 \\ -1 & 6 & 3 & 5 \\ -2 & 6 & 3 & 5 \\ 2 & -4 & -4 & -4 \\ -4 & 1 & 4 & -1 \\ -3 & 6 & 1 & 4 \end{bmatrix}
$$

$$
\text{(f)} \quad \begin{bmatrix} -1 & -4 & 2 & -1 & 3 \\ 0 & -4 & 2 & -1 & 3 \\ -2 & 4 & -5 & -2 & -3 \\ 4 & 3 & -2 & 5 & -2 \\ 5 & 3 & -2 & 5 & -2 \end{bmatrix}
$$

$$
\text{(g)} \quad \begin{bmatrix} 5 & -3 & 2 & -4 & -3 & 4 \\ -4 & -3 & 0 & -3 & -3 & -4 \\ -4 & -3 & 1 & -3 & -3 & -4 \\ 3 & 4 & -2 & 4 & 4 & 2 \end{bmatrix}
$$

5. For each exercise in Exercise 5.5.1, determine the von Neumann value and the optimal strategy.

6. For each of the following games of chance, determine the von Neumann value for the game, and optimal strategies for both players.

(a) Both players secretly flip a coin, they see their own result but not the other. Suppose Heads is greater value than Tails.

Rowan then has two choices. He may *CALL*: both coins are revealed. If Rowan wins, Colleen gives him $2, if Colleen wins, Rowan gives her $4. If both are the same, no money changes hands. He may *BID*: Colleen then has two choices.

Colleen may *FOLD*: and Rowan wins $4. Or she may *SEE*: in which case both coins are revealed, and the winner is awarded $10 from the loser. If there is no winner, no money changes hands.

(b) Each player adds $2 to the pot. Then they roll a 4-sided dice in secret. Each player knows their own results, but not the other's.

Rowan two options. He may *FOLD*: The pot goes to Colleen. He may *PLAY*: In which case he adds $5 to the pot.

Colleen then has two options. She may *FOLD*: The pot goes to Rowan. She may *PLAY*: In which case she adds $5 to the pot.

Then the results are revealed. Whoever wins takes the pot. If they are a tie, then the pot is split between the players.

(c) There are 6 cards, 2 Jacks, 2 Queens and 2 Kings with Jack < Queen < King. The players each place \$1 in the pot. Then, one card each is dealt to each player face down. They may see their own card but not their opponent's.

Rowan now has three choices. He may *FOLD*: the pot goes to Colleen. He may *CHECK*: the pot remains as is, or he may *RAISE*: Rowan adds \$2 to the pot.

If Rowan doesn't FOLD, Colleen also has three choices. She may also *FOLD*: The pot goes to Rowan. She may *SEE*: Colleen adds money to the pot equal to what Rowan added (if any). She may *SEE-RAISE*: Colleen adds money to the pot equal to what Rowan added, and then they both add in an additional \$2.

If no one has folded at this point, the cards are revealed. The pot goes to the winner. If the cards are a tie, then the pot is split evenly between the players.

7. Prove that if a reduced payoff matrix has multiple saddle points, they have the same value.

8. Consider the payoff matrix

$$\begin{bmatrix} x & 0 \\ 0 & y \end{bmatrix}.$$

(a) Find necessary and sufficient conditions for this matrix to be reduced by domination.

(b) Find the von Neumann value in terms of x and y and optimal strategies for each player for the game above. There may be multiple cases.

9. Consider the payoff matrix

$$\begin{bmatrix} -x & x+1 \\ x+1 & -x-2 \end{bmatrix}$$

where $x > 0$.

(a) Find the von Neumann value and optimal strategies for each player for the game above.

(b) Suppose x could be any value, when does this matrix reduce via domination?

(c) Find the von Neumann value and optimal strategies for each player for the game above if x could be any value.

10. Consider the payoff matrix

$$\begin{bmatrix} a & c \\ d & b \end{bmatrix}$$

where $a < b < c < d$.

(a) Find the optimal solutions and the von Neumann value for the above game. (Without loss of generality, we may assume a, b, c, d are positive, why?)

(b) Prove that in a game with a two-by-two payoff matrix where the optimal solution for each player is a *pure strategy*, then the matrix has a saddle point.

(c) Prove via contradiction that in a game with a reduced $n \times n$ payoff matrix where the optimal solution for each player is a *pure strategy*, then the matrix has a saddle point.

Chapter 6

The Transportation & Assignment Problems

Another application of linear optimization is shipping goods from multiple sources to multiple destinations in a way that minimizes costs while satisfying all supply and demand needs. A related task is assigning entities to tasks in a way that minimizes cost or maximizes utility. While these may be modeled as linear optimization problems, we also discuss some algorithms which computes these solutions in a less unwieldy manner.

In Section 6.1, we discuss the general transportation problem and an algorithm which heuristically produces a "good" feasible solution. Then in Section 6.2 we explore an algorithm which improves a feasible solution and produces an optimal solution. Then in Section 6.3, we examine a specific type of transportation problem, the assignment problem, and study an algorithm particular to this class of problem.

6.1 A Transportation problem and VAM

We introduce the transportation problem, consider its connection to linear optimization, and show an algorithm that produces a (maybe not optimal) solution.

Exploration 6.1.1 Suppose we have a company moving goods (let's say widgets) from 3 different warehouses to 3 different markets. The cost of shipping from warehouses to markets is listed below, along with the demand from each market and the supply available at each warehouse:

	Market 1	Market 2	Market 3	
Warehouse 1	\$2/ton	\$1/ton	\$5/ton	70 tons
Warehouse 2	\$5/ton	\$3/ton	\$6/ton	20 tons
Warehouse 3	\$1/ton	\$2/ton	\$8/ton	10 tons
	40 tons	40 tons	20 tons	100 tons

We want to ship the goods from the warehouses to the markets in a way that minimizes costs.

(a) Just eyeballing this, can you come up with a heuristic guess as to an optimal, or at least "good" shipping schedule? How did you come up with this and what did you have to consider?

(b) Let x_{ij} denote the the tons of goods shipped from warehouse i to market j. Write an (in)equality for the quantity of goods shipped from Warehouse 1 in terms of the x_{ij}.

(c) Write an (in)equality for the quantity of goods shipped to Market 2 in terms of the x_{ij}.

(d) Write a (possibly noncanonical) linear minimization problem that minimizes the cost to ship these goods.

Definition 6.1.2 A transportation problem where the total demand and the total supply are the same is a **balanced** transportation problem. $\diamond$

Activity 6.1.3 While we could solve this transportation by the Simplex Algorithm, it would be painfully tedious to do. We develop an algorithm to simplify this process.

(a) Consider the transportation problem described in Exploration 6.1.1, and consider the linear equations

$$\sum x_{i\ell} = w_i, \sum x_{kj} = m_j.$$

Set up an augmented matrix representing this system and solve it. What is the rank of this matrix? Describe the set of solutions to this system.

(b) We now generalize to a balanced transportation problem with n warehouses and m markets with supplies w_i and demands m_j. Note that we have linear equations

$$\sum x_{i\ell} = w_i, \sum x_{kj} = m_j.$$

Since $\sum w_i = \sum m_j$, how many of the x_{ij} are needed to be nonzero for a basic solution?

Hint. Imagine an augmented matrix with the coefficients of the x_{ij} on one side and the supplies/demand of the other. What is an upper bound of the rank of this matrix? Then consider that $\sum w_i = \sum m_j$. What does this say about the (in)dependence of the rows? What then must the rank be? What happened in part **(a)**?

(c) We first mark the difference between the lowest two values in each row/column:

	1	1	1	
1	2	1	5	70
2	5	3	6	20
1	1	2	8	10
	40	40	20	

A table like this is called a **transportation tableau** .

Ideally we would always want to move everything with the cheapest available option. It's not hard to see that in most cases, like this one, this isn't actually possible. What do these numbers we computed measure? How can they help us decide how to move goods?

(d) We select the row or column with the highest difference and use the smallest entry in said row/column to move as much of the goods as we can:

	1	1	1	
1	2	1	5	70
2	5	③	6	20
1	1	2	8	10
	40	40	20	

	1	1	1	
1	2	1	5	70
2	5	③^{20}	6	0
1	1	2	8	10
	40	20	20	

Would it make sense to move any more goods from Warehouse 2? How should we decide how to move goods next?

(e) Note that Warehouse 2 has all their supply exhausted, and shipping from there is no longer an option. What are the differences between the lowest costs and second lowest costs in each row/column?

	?	?	?	
?	2	1	5	70
NA	5	③^{20}	6	0
?	1	2	8	10
	40	20	20	

(f) The next highest difference is for Market 3:

	?	?	?	
?	2	1	⑤	70
NA	5	③^{20}	6	0
?	1	2	8	10
	40	20	20	

	?	?	NA	
?	2	1	⑤^{20}	50
NA	5	③^{20}	6	0
?	1	2	8	10
	40	20	0	

Does it make sense to continue to move goods to Market 3?

What should be the next choice of warehouse/market?

(g) Finish moving the goods from warehouses to markets.

(h) Consider the final transportation tableau:

	NA	NA	NA	
NA	$(2)^{30}$	$(1)^{20}$	$(5)^{20}$	0
NA	5	$(3)^{20}$	6	0
NA	$(1)^{10}$	2	8	0
	0	0	0	

Verify that this is a feasible solution to the original transportation problem. Do you think it is optimal?

(i) Out of the nine possible warehouse/market pairs, how many have actual shipments between them? How does that compare to what we found in (a)?

Definition 6.1.4 To summarize the *Vogel Advanced-Start Method* or *VAM* method of producing a feasible solution to the transportation problem is outlined as follows.

1. We begin with m **sources** (warehouses in the above activities) and n **sinks** (markets in the above activities), each with a supply or demand respectively. We associate each row of the transportation tableau with a source, each column with a sink, and each entry c_{ij} with the cost of shipping from source i to sink j.

2. For each row and column, we record the difference between the lowest two values.

3. We pick the row/column with the largest difference (so long as the associated supply/demand is positive), and the smallest entry in the row/column, c_{ij}. By convention we circle this entry.

4. We "ship" quantity from source i to sink j, recording this quantity as a superscript in the numerator and adjusting the supply for source i and demand for sink j appropriately.

5. We ignore any source/sink with 0 supply/demand and repeat 2-4.

6. If all source/sinks are exhausted and we have not yet circled $m + n - 1$ entries, we circle any entries in the last row/column we've examined, noting that the quantities "shipped" for these entries is zero.

7. Once $m + n - 1$ entries are circled and all supply/demand is exhausted, we terminate. The circled entries are called the **basis** of the tableau.

$\Diamond$

6.2 The Transportation Algorithm

We show how we can take a feasible transportation solution (say from Section 6.1), and from there produce an optimal solution.

Exploration 6.2.1 Consider a transportation tableau:

$$
\begin{array}{cc|c}
\textcircled{5}^{\,20} & \textcircled{5}^{\,30} & 0 \\[4pt]
\textcircled{8}^{\,10} & \boxed{5}^{\,0} & 0 \\ \hline
0 & 0 &
\end{array}
$$

The box here denotes that the bottom right 5 isn't currently being used but likely should be.

(a) It is clear that we should shift some of warehouse 2's shipments to market 2 to reduce costs. Why isn't this a valid transportation tableau?

$$
\begin{array}{cc}
\textcircled{5}^{\,20} & \textcircled{5}^{\,30} \\[4pt]
\textcircled{8}^{\,0} & \boxed{5}^{\,10}
\end{array}
$$

(b) How should we adjust these values to have a valid tableau?

$$
\begin{array}{cc}
\textcircled{5}^{\,?} & \textcircled{5}^{\,?} \\[4pt]
\textcircled{8}^{\,0} & \boxed{5}^{\,10}
\end{array}
$$

(c) How about this one?

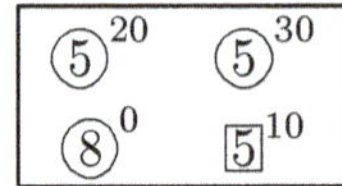

$$
\begin{array}{cccc|c}
\circledast^{\,13} & \circledast^{\,15} & ? & ? & 0 \\[4pt]
\circledast^{\,7} & ? & \circledast^{\,12} & ? & 0 \\[4pt]
? & \circledast^{\,5} & ? & \boxed{\ast}^{\,0} & 0 \\[4pt]
? & ? & ? & ? & 0 \\[4pt]
? & ? & \circledast^{\,9} & \circledast^{\,6} & 0 \\ \hline
0 & 0 & 0 & 0 &
\end{array}
$$

Activity 6.2.2 Recall the transportation tableau obtained in Activity 6.1.3:

118

	NA	NA	NA	
NA	②^{30}	①^{20}	⑤^{20}	0
NA	5	③^{20}	6	0
NA	①^{10}	2	8	0
	0	0	0	

Note that this tableau corresponds to a basic feasible solution for the original minimization problem.

(a) If this solution is not optimal, what should the next step be? (Recall how we chose pivots in Activity 2.2.4.)

    A. Pick a variable to exit the basis which increases the objective function and make the smallest change.

    B. Pick a variable to exit the basis which decreases the objective function and make the smallest change.

    C. Pick a variable to exit the basis which increases the objective function and make the biggest change.

    D. Pick a variable to exit the basis which decreases the objective function and make the biggest change.

(b) Select row values a_i and column values b_j so that each *circled* value c_{ij} is the sum of row and column values $a_i + b_j$.

(We can think of these as analogous to the shadow costs associated with the warehouse/market bounds.)

	b_1	b_2	b_3
a_1	②^{30}	①^{20}	⑤^{20}
a_2	5	③^{20}	6
a_3	①^{10}	2	8

(c) Replace each entry of the tableau c_{ij} with $c_{ij} - a_i - b_j$. What does this measure? What does it mean if each entry is nonnegative?

(d) Pick a $c_{ij} < 0$.

	b_1	b_2	b_3
a_1	⓪^{30}	⓪^{20}	⓪^{20}
a_2	?	⓪^{20}	$\boxed{-?}$
a_3	㉠^{10}(?)	?	?

(e) Pick circled entries $c_{k\ell}$ so that they with the boxed c_{ij} form a **cycle**, that is each of these entries shares a row with exactly one another of the entries, and a column with another of the entries.

(f) Based on the discussion in **(a)**, which entry should transfer their shipments to c_{23}?

A. c_{11}.   D. c_{22}.

B. c_{12}.

C. c_{13}.   E. c_{31}.

How do the other entries in the cycle adjust? (There may be more than one valid choice.)

(g) Remove the basis entry which is no longer being used, and recompute the a_i, b_j with the new basis.

	b_1	b_2	b_3
a_1	?	?	?
a_2	?	?	?
a_3	?	?	?

(h) Verify that none of the entries are non negative. Why do we now terminate?

We then replace the entries with the original entries:

	b_1	b_2	b_3
a_1	2^{30}	1^{40}	5^{0}
a_2	5^{0}	3^{0}	6^{20}
a_3	1^{10}	2^{0}	8^{0}

(i) Use Sage to confirm the solution:

```
%display typeset
A = (FIXME)
b = (FIXME)
c = (FIXME)
P = InteractiveLPProblem(A, b, c,
  constraint_type = "==",
  variable_type = ">=",
  problem_type = "min")
P
```

```
print(P.optimal_solution())
print(P.optimal_value())
```

Definition 6.2.3 To summarize, the Transportation Algorithm is as follows:

1. We begin with a feasible transportation tableau, probably via **VAM** Definition 6.1.4.

    We then associate with each row an unknown a_i and each column a b_j.

2. We select values a_i, b_j so that for each basis entry c_{ij} we have that $c_{ij} = a_i + b_j$.

3. Replace every entry c_{ij} with $c_{ij} - a_i - b_j$.

4. If there is a negative entry $c_{k\ell}$ box this entry and select basis entries so that they, along with the boxed entry, form a cycle.

   If each entry is nonnegative, GOTO 6.

5. Shift shipments appropriately along the cycle so that $c_{k\ell}$ has a nonnegative shipping quantity, and one of the selected basis entries has a zero shipping quantity. Remove this entry from the basis and add the entry in step 4 to the basis.

   Then GOTO 2.

6. Replace each cost entry with the costs from step 1 and terminate.

$\Diamond$

Activity 6.2.4 Consider the tableau:

	b_1	b_2	b_3
a_1	③¹	④⁵	6
a_2	①³	3	1
a_3	②⁰	2	②²

(a) Perform steps 2-5 of the Transportation Algorithm Definition 6.2.3. How much was shifted along the cycle in step 5?

(b) What is the analogous type of pivot when performing the Simplex Algorithm Definition 2.2.6?

Activity 6.2.5 Consider the Transportation Algorithm Definition 6.2.3. Recall that the objective function for the transportation problem is $f(\mathbf{x}) = \sum x_{ij} c_{ij}$, and that the entries of the tableau produced in step 3 are $c_{ij} - a_i - b_j$.

Show that in step 5 the newly produced solution has a lower or equal objective value.

6.2.1 Unbalanced Transportation Problems

Activity 6.2.6 We now consider the case of **unbalanced** transportation problems, where the demand and supply are unequal.

(a) Suppose we had the following transportation problem:

	Market 1	Market 2	Market 3	
Warehouse 1	$5/ton	$3/ton	$1/ton	35 tons
Warehouse 2	$6/ton	$2/ton	$5/ton	45 tons
Warehouse 3	$4/ton	$2/ton	$1/ton	15 tons
	30 tons	20 tons	40 tons	

Suppose we satisfy all the demand in a way that minimizes costs. What would be the remaining result?

(b) Suppose we "ship" the excess supply to a phantom "market":

	Market 1	Market 2	Market 3	"Market"	
Warehouse 1	$5/ton	$3/ton	$1/ton	?	35 tons
Warehouse 2	$6/ton	$2/ton	$5/ton	?	45 tons
Warehouse 3	$4/ton	$2/ton	$1/ton	?	15 tons
	30 tons	20 tons	40 tons	?	

How much is shipped to the "market"? How much does it cost to "ship" from each warehouse to "market"?

(c) Suppose we had the following transportation problem:

	Market 1	Market 2	Market 3	
Warehouse 1	$5/ton	$3/ton	$1/ton	35 tons
Warehouse 2	$6/ton	$2/ton	$5/ton	45 tons
Warehouse 3	$4/ton	$2/ton	$1/ton	15 tons
	30 tons	50 tons	40 tons	

Suppose we exhaust all the supply in a way that minimizes costs. What would be the remaining result?

(d) Suppose we have a phantom "warehouse" that "filled" the outstanding demand.

	Market 1	Market 2	Market 3	
Warehouse 1	$5/ton	$3/ton	$1/ton	35 tons
Warehouse 2	$6/ton	$2/ton	$5/ton	45 tons
Warehouse 3	$4/ton	$2/ton	$1/ton	15 tons
"Warehouse"	?	?	?	?
	30 tons	50 tons	40 tons	

How much additional "supply" is needed? How much would it cost to ship this "supply" from "warehouse" to the markets?

(e) Describe a general procedure for solving unbalanced transportation problems.

6.3 The Assignment Problem and Hungarian Algorithm

We consider the assignment problem, where each source and sink have supply and demand 1, and an alternative algorithm to solve these sort of problems.

Exploration 6.3.1 Suppose we have 3 jobs and 3 contractors, and we wish to assign jobs to contractors at the minimum price. How can we distribute the jobs amongst the contractors? (Costs are in thousands of dollars.)

	Job 1	Job 2	Job 3	
Contractor 1	10	9	12	?
Contractor 2	9	9	10	?
Contractor 3	10	7	9	?
	?	?	?	?

(a) What are the "supply" and "demand" of each job and contractor?

(b) Use VAM Definition 6.1.4 and the Transportation Algorithm Definition 6.2.3 to solve this problem.

(c) Could there have been an easier way to approach this problem?

Definition 6.3.2 An **assignment problem** is a transportation problem where the supply and demands are all 1. $\Diamond$

Note that in Exploration 6.3.1, while we were able to solve this as a transportation problem, the restriction to supplies and demands of 1 ought to yield a simpler way to find a solution.

Definition 6.3.3 Let T be a tableau for a balanced assignment problem. A **permutation set of zeroes** is a subset of zero cells for T so that each row and column contains exactly one zero cell. $\Diamond$

Activity 6.3.4 We explore some features of the assignment tableau that can help shed light on what an appropriate algorithm would look like.

a_{11}	a_{12}	a_{13}
a_{21}	a_{22}	a_{23}
a_{31}	a_{32}	a_{33}

(a) If we multiply each entry by 2, would that change our optimal solution? What about by -1? What values k could we multiply the tableau by to preserve the optimal solution?

(b) If we add or subtract 2 from each entry, does it change the optimal solution? What about n?

(c) Suppose we had an optimal solution to the assignment problem. Explain why adding k to each entry in a row does not change the optimal solution.

    Hint. How would this compare to solving the original problem and adding k to the cost?

(d) What would change if we did this to a different row? A column?

(e) Suppose we subtracted the smallest value in each row from every entry of that row. If there was a permutation set of zeroes, what would that entail?

(f) Suppose we subtracted the smallest value in each column from every entry of that column. If there was a permutation set of zeroes, what would that entail?

(g) If we had a tableau with all rational values, how could we change this to a tableau of all integer values with the same optimal solution?

(h) If we had a tableau with all integer values, how could we change this to a tableau of all nonnegative integer values with the same optimal solution?

Definition 6.3.5 We state here the steps of the *Hungarian Algorithm*. We start with an $n \times n$ assignment tableau T.

1. IF the entries of T are rational but not all integral, $a_{ij} = \frac{p_{ij}}{q_{ij}}, q_{ij} > 0$:

   THEN multiply each entry of T by the lowest common multiple of the denominators, $\operatorname*{lcm}_{i,j} q_{ij}$.

2. IF the entries of T are not all nonnegative:

   THEN add to every entry of T the smallest value of T, $\min\limits_{i,j} a_{ij}$.

3. Subtract from each row the smallest entry in that row.

4. Subtract from each column the smallest entry in that column.

5. IF T has a permutation set of zeroes: STOP.

6. Draw a **minimum*** number of lines through T covering an entire row or column such that all 0's are covered.

7. Let M be the value of the smallest uncovered entry. Subtract all uncovered entries by M, and add M to the entries corresponding to intersections of the lines.

8. GOTO 5.

$\Diamond$

Activity 6.3.6

(a) In step 6 of Definition 6.3.5, suppose we draw k lines. Could k be greater than n?

(b) Suppose that the zeroes of the tableau in a given step do *not* form a permutation set of zeroes. Show that $k < n$.

(c) What would it mean if $k = n$?

(d) Show that step 7 is equivalent to subtracting M from each uncovered row, and adding M to each covered column. (Or subtracting M from each uncovered column, and adding M to each covered row.)

(e) Why does step 7 not change the optimal assignment?

Activity 6.3.7 Use the Hungarian Algorithm Definition 6.3.5 to solve the problem from Exploration 6.3.1.

Activity 6.3.8 Suppose 4 students are picking 4 research topics. The four topics are to be distributed one each amongst the four students. They rated the topics on a scale of 1-10.

	Topic 1	Topic 2	Topic 3	Topic 4
Student 1	6	9	10	8
Student 2	2	8	9	7
Student 3	7	7	8	9
Student 4	6	8	9	8

We want to maximize the total "enjoyment".

(a) This is a maximization problem, and the assignment problem is a minimization problem, how might we convert it to a minimization problem?

(b) After converting, use the Hungarian Algorithm Definition 6.3.5 to solve the problem.

6.4 Summary of Chapter 6

We introduce the **Transportion Problem**: given n "warehouses" and m "markets", each with W_i supply and M_j demand respectively, then given c_{ij}, the cost to ship from warehouse i to market j, to find the shipping quantities x_{ij} which satisfy all the warehouse and market constraints. When $\sum_{i=1}^{n} W_i = \sum_{j=1}^{m} M_j$, we call this problem **balanced**.

This can be captured with a transportation tableau:

$$
\begin{array}{cccc|c}
c_{11} & c_{12} & \cdots & c_{1m} & W_1 \\
c_{21} & c_{22} & \cdots & c_{2m} & W_2 \\
\vdots & \vdots & \ddots & \cdots & \cdots \\
c_{n1} & c_{n2} & \cdots & c_{nm} & W_n \\
\hline
M_1 & M_2 & \cdots & M_m & \sum_{i=1}^{n} W_i = \sum_{j=1}^{m} M_j
\end{array}
$$

Note that if we were to consider the inherent system of equations:

$$\sum_{j=1}^{m} x_{ij} = W_i$$

$$\sum_{i=1}^{n} x_{ij} = M_j$$

and the fact that $\sum_{i=1}^{n} W_i = \sum_{j=1}^{m} M_j$, that the associated augmented matrix would have rank $n + m - 1$ and thus at most $n + m - 1$ of the x_{ij} need be nonzero for a feasible or optimal solution. A selection of these variables will be considered the **basis** of a solution, and is equivalent to the basis variables from Chapter 2.

We then introduce the Vogel Advanced Start Method (VAM) Definition 6.1.4 to heuristically pick a "good" feasible solution. The essential premise is that, we take each warehouse and market, and consider the difference between the cheapest and second cheapest options for that row/column. Since we want to minimize cost, we prioritize warehouses/markets where making the second best choice would incur a larger cost penalty than the best choice, and choose the cheapest option for those row/columns. We then readjust and repeat until we obtain a feasible solution.

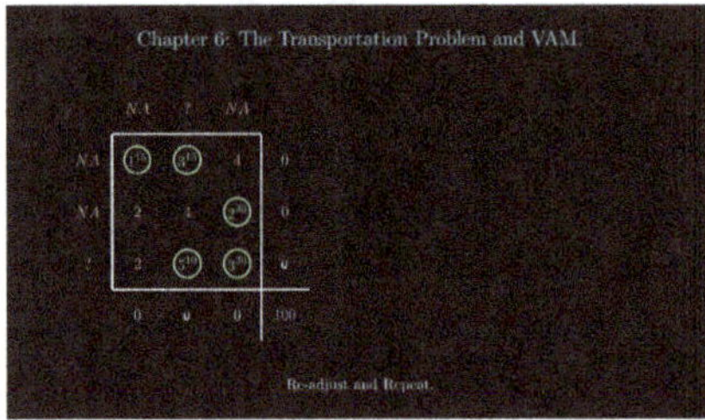

Figure 6.4.1 The Vogel Advanced Start Method.

We then proceed with the **Transportation Algorithm** Definition 6.2.3. The general idea is that we assign an a_i to each row and b_j for each column so that $c_{ij} = a_i + b_j$ for each x_{ij} in the basis. Then, we reduce each c_{ij} by $a_i + b_j$. We then see if any entries are negative.

Note that for the current basis, the cost of shipping is $\sum_{i=1}^{n} \sum_{j=1}^{m} c_{ij} x_{ij}$, but since $c_{ij} = 0$ for x_{ij} not in the basis, and since

$$\sum_{j=1}^{m} x_{ij} = W_i \text{ and } \sum_{i=1}^{n} x_{ij} = M_j,$$

we have that

$$\sum_{i=1}^{n} \sum_{j=1}^{m} c_{ij} x_{ij} = \sum_{i=1}^{n} a_i W_i + \sum_{j=1}^{m} b_j M_j,$$

so shifting the shipping to an entry where $c_{ij} < a_i + b_j$ would lower the total shipping cost. We then outline a cycle consisting of new basis elements and shifting the shipping around to preserve the warehouse and market constraints, adding the negative entry to the basis and removing an entry with no shipping from the basis. Note that it is possible for this shift to be zero, yet changing the basis, equivalent to a degenerate pivot from Section 2.3. We repeat this process until there are no longer negative entries.

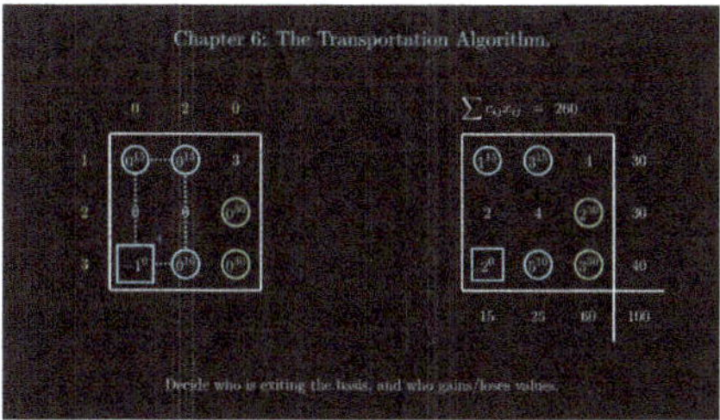

Standalone

Figure 6.4.2 The Transportation Algorithm.

Finally, we consider the **Assignment Problem** and the **Hungarian Algorithm** Definition 6.3.5. The assignment problem can be thought of as a transportation problem where each warehouse and market have supply and demand 1. But since this is greatly simplified, we should expect a simpler algorithm. We note that we may add and subtract k from any row or column without changing the optimal assignment, since this is equivalent to picking originally and then giving/taking k back afterwards. We may also multiply all entries by the same value and preserve the optimal assignment by a similar argument. So we may adjust the tableau to only have nonnegative integer entries, and then subtract the smallest value from each row, then each column.

If there is a **permutation set of zeroes**, a collection of n zeroes no two of which share a row or column, then these clearly represent an optimal assignment. If none exist, we may rearrange the tableau by drawing a minimum number of lines through each zero. We then take the smallest uncovered entry and subtract that from each uncovered row and add it to each covered column. Equivalently we subtract this value from each uncovered entry and add it to each intersection. We then repeat until we find a permutation set of zeroes.

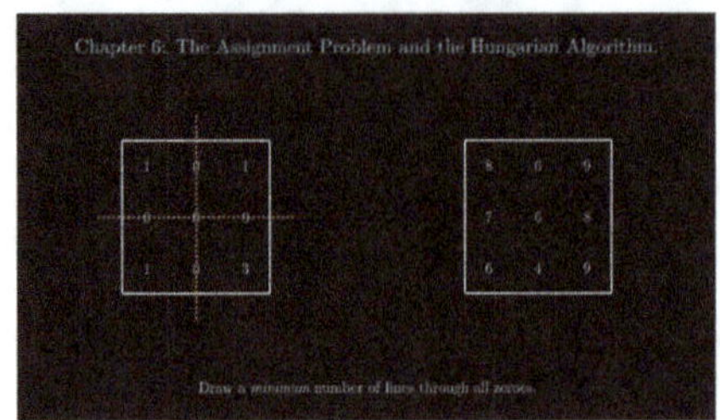 

Figure 6.4.3 The Assignment Problem.

6.5 Problems for Chapter 6

Exercises

1. Solve each of the following transportation problems.

(a)

7	4	7	5
4	2	6	20
5	4	6	10
15	10	10	

(b)

4	3	7	10
2	1	4	20
3	2	4	10
15	20	5	

(c)

2	4	1	25
4	5	4	5
5	10	15	

(d)

5	9	5
6	7	10
7	8	25
30	10	

(e)

4	7	3
5	8	2
9	10	6
10	8	7
6	12	

(f)

1	7	10	10	5
4	1	2	3	16
2	3	4	6	3
5	9	4	6	

(g)

3	5	7	14
5	4	7	7
6	7	11	5
8	9	12	4
5	7	18	

(h)

5	7	9	7	8
6	8	5	4	15
4	5	6	7	8
6	6	8	5	11
10	11	8	13	

(i)

4	4	6	4	5
4	5	6	4	14
6	7	7	6	4
7	8	8	7	6
8	6	10	5	

2. In Exercise 6.5.1, looking at the final tableau of the Transportation Algorithm, determine which of these have a *unique* optimal solution.

3. In Exercise 6.5.1, for each problem *without* a unique optimal solution, find an alternative solution and show that it is feasible and gives the same objective value.

4. *TRUE or FALSE* for a valid choice of a_i, b_j in Definition 6.2.3 step 2, then for a fixed real number Z, $a_i - Z, b_j + Z$ is also a valid choice.

5. Suppose Z were an absurdly large number. Solve the transportation problem:

3	4	Z	8
4	7	4	6
Z	4	4	7
5	4	3	10
10	9	12	

6. Solve the following transportation problems.

4	6	5	6	8
6	6	4	7	19
6	5	4	5	14
4	6	3	7	14
15	20	20	10	

3	5	6	5	15
3	4	3	5	5
3	3	3	5	15
3	6	5	5	10
10	7	9	12	

7. On the planet Zeltros, a luxury resort is preparing for a multi-day celebration they are hosting. One of the requested activities by the guests is hoverbike riding. Over the three day event, they will require 500 hoverbikes on the first day, 300 hoverbikes on the second and 800 hoverbikes on the final day.

   Brand new hoverbikes are 1000 credits. After use they undergo a maintenance procedure which costs 250 credits and takes two days, before they may be used again. A rush job that takes one day is possible but costs 400 credits.

   The resort naturally wishes to minimize their costs. They wish to know how many bikes to purchase and what maintenance schedule would achieve this.

   Model this as a transportation problem where the "warehouses" are the number of brand new hoverbikes purchased before day 1, used hoverbikes from day 1 and used hoverbikes from day 2. Can we make educated guesses on what these capacities should be? What should the "markets" represent and what are their demands? Then solve the problem.

8. Solve the following assignment problem using both Definition 6.3.5 and as a transportation problem using Definition 6.2.3.

2	1	2
5	4	6
1	2	5

9. Solve the following assignment problems.

(a)

33	23	22
38	27	30
38	33	34

(b)

$$\begin{array}{cccc} 2 & 5 & 6 & 4 \\ 3 & 7 & 5 & 3 \\ 3 & 5 & 3 & 4 \\ 4 & 5 & 5 & 5 \end{array}$$

(c)

$$\begin{array}{cccc} 11 & 13 & 15 & 18 \\ 9 & 8 & 12 & 16 \\ 15 & 16 & 17 & 20 \\ 14 & 14 & 16 & 19 \end{array}$$

(d)

$$\begin{array}{ccccc} 12 & 6 & 9 & 11 & 7 \\ 14 & 7 & 8 & 12 & 9 \\ 17 & 11 & 12 & 14 & 11 \\ 18 & 10 & 13 & 17 & 12 \\ 13 & 6 & 8 & 11 & 6 \end{array}$$

(e)

$$\begin{array}{ccccc} 13 & 10 & 13 & 10 & 15 \\ 13 & 8 & 11 & 10 & 13 \\ 9 & 4 & 9 & 4 & 10 \\ 10 & 4 & 8 & 6 & 111 \\ 14 & 7 & 12 & 9 & 13 \end{array}$$

10. Some of the problems in Exercise 6.5.9 have multiple solutions. Identify them and identify all the optimal solutions.

11. Consider a general balanced transportation problem. Let $w_1, \ldots, w_n$ denote the supply of the n warehouses, and $m_1, \ldots, m_m$ denote the demand of the m markets. Then let c_{ij} denote the cost to ship from warehouse i to market j.

   (a) Write out a noncanonical minimization problem which models this problem (there should be some equality constraints).

   (b) State the dual problem to the transportation problem.

   (c) Prove that in Definition 6.2.3, for a valid choice of a_i, b_j in step 2 that

   $$f(\mathbf{x}) = \sum_i a_i w_i + \sum_j b_j m_j$$

   How does this relate the dual problem to the transportation problem?

   (d) Consider the nonnegativity condition in step 4 of Definition 6.2.3. How does this relate to the constraints of the dual problem to the transportation problem?

12. Consider an $n \times n$ assignment problem treated as a transportation problem. Apply VAM (Definition 6.1.4) to this initial tableau. Show that the resulting basis generated has $n - 1$ selected entries where the shipment along those entries (x_{ij}) is zero.

13. Prove that the Transportation Algorithm Definition 6.2.3 will lead to an optimal solution after a finite number of steps if the algorithm is applied to a problem with all data rational.

14. Consider the following network:

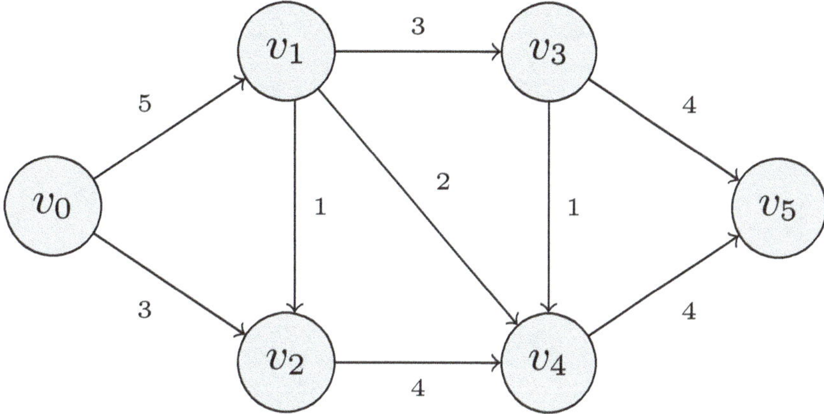

where the labels on each edge denote the "distance" from v_i to v_j.

(More about directed weighted networks may be found in Definition 7.3.2)

(a) Consider an "assignment" problem with "sources" and "sinks" $v_0, \ldots, v_5$ such that the "cost" of from shipping from v_i to v_j is the label on the $v_i v_j$ edge. Let the cost from shipping from v_5 to v_0 be 0. For each $i, 1 \le i \le 4$, let the cost of shipping from v_i to itself be 0. Let every other cost be an absurdly high Z.

Solve this assignment problem.

(b) For your optimal solution, if v_i "ships" to v_j, highlight the corresponding edge on the network above. If a vertex v_k ships to itself, highlight that vertex. Disregard the case when v_5 ships to v_0. What can you say about the highlighted edges? What may be reasonable conjectures about these edges?

15. Consider a general directed weighted network with vertices $v_0, \ldots, v_n$ with a unique source v_0 and sink v_n, and all weights on edges are positive such as Exercise 6.5.14 (see Definition 7.3.2 for more information).

We will be considering assignment problems where "warehouses" and "markets" are the vertices $v_0, \ldots, v_n$. Throughout this problem, let Z be an absurdly large number.

(a) Consider an assignment problem where the cost of shipping from v_i to v_j is the weight on that edge if it exists and Z otherwise. Show that any solution to the assignment problem (not necessarily optimal) is bijective with the set of possible unions of disjoint cycles on the network.

(b) Suppose we adjust the previous assignment problem where the cost of shipping from v_n to v_0 is 0. Show that solutions to this assignment problem are bijective with the union of disjoint cycles and path from v_0 to v_n on this network.

(c) Suppose we then further adjust this assignment problem by setting the cost to ship from v_i to itself for any $1 \leq i \leq n - 1$ to be 0. Show that an *optimal* solution to this problem must be a shortest path from v_0 to v_n.

Chapter 7

Network Flows

In this chapter, we focus on flows through directed graphs called *networks*. While graph theory is an enormous area of math (in which I work), we are particularly interested in optimization problems on weighted and or capacitated networks, such as maximizing the flow through a network with limited capacity, or finding a path minimizing the distance between nodes.

In Section 7.1 we explore maximal flows through capacitated networks and model this problem with linear optimization. In Section 7.2 we discuss the dual problem, connect it to *cuts* in the network, and discuss algorithms which solve both problems. In Section 7.3 we shift our attention to weighted networks, and examine shortest paths and minimum cost flows.

7.1 Directed Graphs and Network Flow

In this section, we introduce capacitated networks and flows.

Exploration 7.1.1 Consider a series of islands with bridges between them. A group of people are trying to move from island v_0 to island v_5. Due to the length/width of the bridges, only but so many people can move between a pair of islands in a minute, and these are labeled. The bridges, represented by arrows below, also only allow movement in one direction. Some pairs of islands have two bridges between them, one going in each direction.

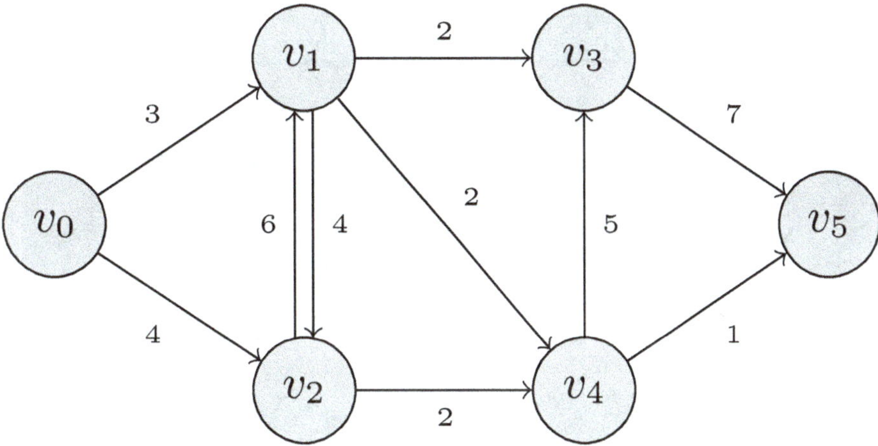

(a) Conjecture a solution to the maximum number of people that can arrive at v_5 in a minute.

(b) If you could increase the capacity of a single bridge to increase the number of people who can travel to v_5 in a minute, which would it be and by how much?

Definition 7.1.2 A **directed graph** or **network** is a pair $D = (V, E)$ where V is a set of **vertices** and E is a set of ordered pairs of distinct elements of V called **edges**.

A network is **capacitated** if for each edge $(v_i, v_j) \in E$ we assign a nonnegative capacity c_{ij}. (If there is no edge from v_i to v_j, we may equivalently say that $c_{ij} = 0$.)

A **flow** assigned to a capacitated network is an assignment to each edge $(v_i, v_j) \in E$ a value x_{ij} such that $0 \leq x_{ij} \leq c_{ij}$. (If there is no edge from v_i to v_j, what must x_{ij} be?)   ◊

Note 7.1.3 Graph theory is a rich, complex and deep area of study. Graph Theorists study a variety of graphs or objects called graphs, with a wide range of conventions. For the purposes of this chapter, graphs are directed, there is at most two edges between graphs (one in each direction), and loops are disallowed. Note that in general some or any of these conventions can be modified or removed.

Definition 7.1.4 For any vertex v_j, the **net input flow** at vertex v_j is $\varphi(v_j) =$
$$\sum_i x_{ij} - \sum_i x_{ji}$$
If $\varphi(v_j) < 0$ then we say v_j is called a **source**.
If $\varphi(v_j) > 0$ then we say v_j is called a **sink**.
If $\varphi(v_j) = 0$ then we say v_j is called an **intermediary vertex**.   ◊

Activity 7.1.5 For the network in Exploration 7.1.1, find three different flows, including one you believe is an "optimal" flow.

(a) For each flow you found, identify the sources, sinks, and intermediary vertices.

(b) For each flow you found, compute the sum $\sum_j \varphi(v_j)$.

Theorem 7.1.6   Conservation of Flow. *For a capacitated network N,* $\sum_i \varphi(v_i) = 0$ *for any flow on N.*

Activity 7.1.7 Prove Theorem 7.1.6.

Activity 7.1.8 Consider the network:

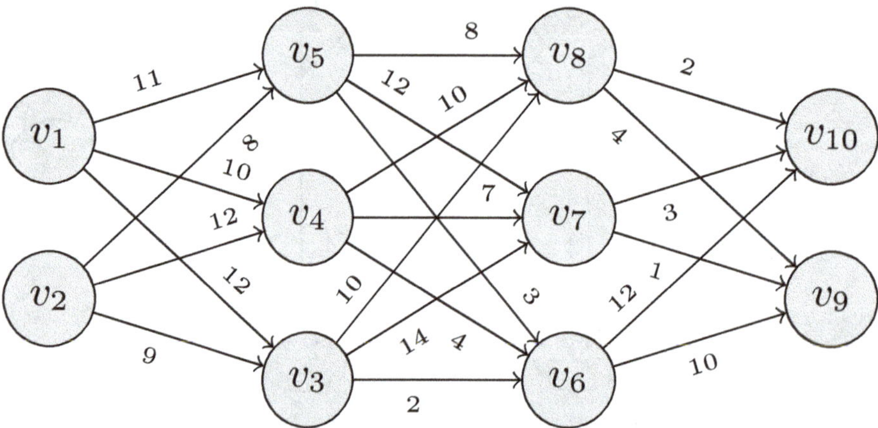

(a) Find a (not necessarily optimal!) flow through this network with exactly two sources and exactly two sinks.

(b) Add two vertices to this network: v_s, v_t, and two edges from v_s to two vertices, and two edges to v_t from two different vertices, each with infinite capacity, and extend the above flow to those edges so that this flow has a unique source and sink.

Observation 7.1.9 To be able to address the sort of questions we wish to ask, we will restrict ourselves to *networks with a unique fixed source or sink, with no edges from the sink or to the source.* In light of Activity 7.1.8, this is not really much of a restriction.

Activity 7.1.10 Suppose we have a capacitated network with a unique fixed source v_s and sink v_d. We wish to define a maximization linear optimization problem with decision variables $x_{ij}, v_i, v_j \in V$.

(a) Define the objective function both in terms of variables x_{sj} and x_{id}. Explain why these are equivalent (can you prove it?).

(b) For each edge (v_i, v_j), there is a natural inequality constraint for the decision variables associated with this edge. What is this inequality?

(c) For each vertex v_j not our source or sink, there is an equality constraint for the decision variables associated with this vertex. Which of the following represents this equality?

A. $\displaystyle\sum_i x_{ji} - \sum_i x_{sj} = 0.$ $\qquad$ D. $\displaystyle\sum_i x_{ji} = 0.$

B. $\displaystyle\sum_i x_{ij} - \sum_i x_{jd} = 0.$ $\qquad$ E. $\displaystyle\sum_i x_{ij} - \sum_i x_{ji} = 0.$

C. $\displaystyle\sum_i x_{ij} = 0.$ $\qquad$ F. $\displaystyle\sum_i x_{ij} + \sum_i x_{ji} = 0.$

(d) There is an additional type of inequality for this problem, what is it?

(e) Write out the primal max problem for Exploration 7.1.1 as a noncanonical Tucker tableau.

(f) Solve this problem:

```
%display typeset
A = (FIXME)
b = (FIXME)
c = (FIXME)
P = InteractiveLPProblem(A, b, c,
  [FIXME],
  constraint_type = [FIXME],
  variable_type = [FIXME],
  problem_type = FIXME)
P
```

```
print(P.optimal_solution())
print(P.optimal_value())
```

(g) Let μ_i denote the dual variable for Exploration 7.1.1 associated with vertex i and let y_{ij} denote the dual variable associated with edge (v_i, v_j).

Describe the dual min problem.

7.2 Max Flow - Min Cut

In this section, we consider an algorithm which solves for the maximum flow. We also discuss *cuts*, and show how duality connects flows and cuts.

Exploration 7.2.1 Recall Exploration 7.1.1. Suppose that we wish to install a toll booth on these bridges so that each person going to v_5 pays a toll at least once. The cost of installing a toll booth on a bridge is proportional to it's capacity. Find three different ways to install these booths, and find what you believe is the cheapest way to do so.

Definition 7.2.2 Given a capacitated network N, a **cut** of N is a partition of the vertex set into nonempty subsets $C = (V_1, V_2)$, where $V = V_1 \sqcup V_2, v_s \in V_1, v_d \in V_2$.

The **capacity** of a cut C is the sum $\displaystyle\sum_{v_i \in V_1, v_j \in V_2} c_{ij}.$ $\Diamond$

Activity 7.2.3

(a) From Exploration 7.1.1, find three different cuts and their capacities.

(b) What cut do you think minimizes the capacity, how does this compare to your conjectured max flow for this problem?

Activity 7.2.4 Consider the primal maximization problem for a max flow problem for a capacitated network with unique source v_s and unique sink v_d:

$$\textbf{Maximize: } \sum_i x_{id}$$

$$\textbf{subject to: } \sum_{v_i \in V} x_{ij} - \sum_{v_i \in V} x_{ji} = 0, \text{ for each non source/sink vertex } v_j$$

$$x_{ij} \leq c_{ij}, \text{ for each edge}(v_i, v_j)$$

$$x_{ij} \geq 0, \text{ for each edge}(v_i, v_j).$$

Go through this activity twice. First for the network and max flow f' found for Exploration 7.1.1, then for a general capacitated network.

(a) Consider the dual problem for this maximization problem where μ_j is the unconstrained dual variable for the vertex equality constraint and the y_{ij} is the dual variable for the capacity constraint. Verify that this problem may be written as

$$\textbf{Minimize: } \sum_{(v_i, v_j) \in E} c_{ij} y_{ij}$$

$$\textbf{subject to: } \mu_j + y_{sj} \geq 0, \text{ for each edge } (v_s, v_j)$$

$$-\mu_i + \mu_j + y_{ij} \geq 0, \text{ for each edge}(v_i, v_j), i \neq s, j \neq d$$

$$-\mu_i + y_{id} \geq 1, \text{ for each edge } (v_i, v_d)$$

$$y_{ij} \geq 0, \text{ for each edge}(v_i, v_j)$$

(b) Verify that we may simplify the dual solution as:

$$\textbf{Minimize: } \sum_{(v_i, v_j) \in E} c_{ij} y_{ij}$$

$$\textbf{subject to:} \quad -\mu_i + \mu_j + y_{ij} \geq 0, \quad \text{for each edge}(v_i, v_j)$$
$$\mu_s = 0$$
$$\mu_d = -1$$
$$y_{ij} \geq 0, \quad \text{for each edge}(v_i, v_j).$$

(c) Suppose $\mu_j = -1$. What could μ_i, y_{ij} be? What value for μ_i would minimize the dual objective function? What happens if μ_i is huge? How would the resulting μ_i that affect the inequality $-\mu_k + \mu_i + y_{ki} \geq 0$?

Repeat for $\mu_j = 0$, and if μ_i were either 0 or -1.

(d) Suppose each $\mu_k \in \{0, -1\}$, Note that $V_1 := \{v_i : \mu_i = 0\}, V_2 := \{v_j : \mu_j = -1\}$ forms a cut of N.

For $v_i, v_j \in V_1$, what is the y_{ij} that minimizes the objective function?

For $v_i, v_j \in V_2$, what is the y_{ij} that minimizes the objective function?

For $v_i \in V_1, v_j \in V_2$, what is the y_{ij} that minimizes the objective function?

For $v_i \in V_2, v_j \in V_1$, what is the y_{ij} that minimizes the objective function?

Can any cut of N be modeled this way?

(e) What is the capacity of the above cut? How does that relate to the dual problem?

(f) Prove that the maximum flow through a network is bounded above by the minimum cut capacity.

Activity 7.2.5 We explore a way of generating potential minimum cuts using a maximum flow. Again recall Exploration 7.1.1 and your proposed maximum flow f'.

(a) Let $v_0 \in V_1'$, we recursively define V_1' by adding v_j to V_1' if either:

- $v_i \in V_1', (v_i, v_j) \in E, x'_{ij} < c_{ij}$.
- $v_i \in V_1', (v_j, v_i) \in E, x'_{ji} > 0$.

(b) Let $V_2' := V \backslash V_1'$. What is the cut capacity of (V_1', V_2')?

Activity 7.2.6 We now prove that the minimum cut capacity is equal to the maximum flow.

(a) Why does the primal max problem achieve optimality?

Call the maximum flow f', with flow on each edge x'_{ij}.

(b) Let $v_s \in V_1'$, we recursively define V_1' by adding v_j to V_1' if either:

- $v_i \in V_1', (v_i, v_j) \in E, x'_{ij} < c_{ij}$.
- $v_i \in V_1', (v_j, v_i) \in E, x'_{ji} > 0$.

and repeating until we stabilize, why must we eventually stabilize?

(c) Show that for any v_k in V_1', there is an α-**path** P: a sequence of vertices starting v_s to v_k, where between v_i and v_j either (v_i, v_j) or $(v_j, v_i) \in E$.

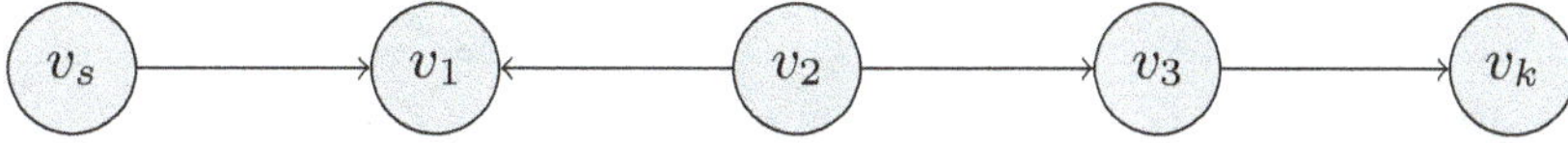

We would call the edges (v_i, v_j) to be **forward edges** and (v_j, v_i) **backwards edges** of P.

(d) Suppose (by way of contradiction) that $v_d \in V_1'$. There is by (c) an α-path P from v_s to v_d.

Let

$$q := \min\left\{ \min_{(v_i,v_j) \text{ a forward edge}} \{c_{ij} - x_{ij}'\}, \min_{(v_j,v_i) \text{ a backwards edge}} \{x_{ji}'\} \right\}.$$

Why is $q > 0$?

(e) We define a new flow f'' as follows: for each forward edge of P, (v_i, v_j), we add $x_{ij}'' = q + x_{ij}'$. For each backwards edge (v_j, v_i) we subtract $x_{ji}'' = x_{ji}' - q$.

Explain why this is still a valid network flow.

(f) Explain why f'' has a greater value than f'. Why must $v_d \notin V_1'$?

(g) Define $V_2' = V \backslash V_1'$. Prove that for any $v_i \in V_1', v_j \in V_2'$, we have that $x_{ij}' = c_{ij}, x_{ji}'' = 0$.

(Not necessary for this proof, but to tie things in, if $x_{ij}'' = c_{ij}$, what does that say about y_{ij} from the dual problem in Activity 7.2.4? If $x_{ij}' < c_{ij}$, what does that say about y_{ij}?)

(h) Use the result from Exercise 7.5.5 to show that the value of f' is equal to the cut capacity of (V_1', V_2'). (Proving the result!)

(i) Going back to Activity 7.2.4 if we let $\mu_i = 0$ for $v_i \in V_1'$ and $\mu_j = -1$ for $v_j \in V_2'$, what is the value of the dual min objective?

Theorem 7.2.7 Max Flow-Min Cut. *Let $N = (V, E)$ be a capacitated directed network with unique fixed source and unique fixed sink, no edges into the source, and no edges out of the sink. Then the value of the maximal flow from v_s to v_d is equal to the minimal cut capacity in N.*

Activity 7.2.8 Suppose we had a nonoptimal flow f, how could we adopt the procedure in Activity 7.2.6 to obtain a better flow?

7.2.1 Algorithms for Max Flow and Min Cut

Exploration 7.2.9 Consider the following capacitated network with source v_0 and sink v_5:

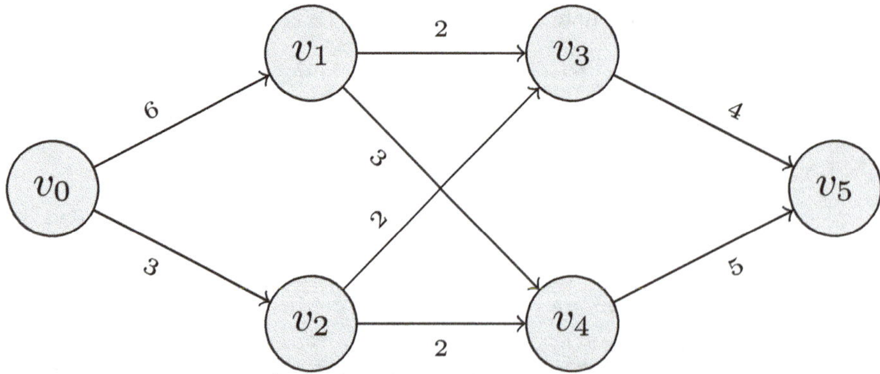

Recall the procedure to produce improved flows in Activity 7.2.6.

(a) Begin with the zero-flow.

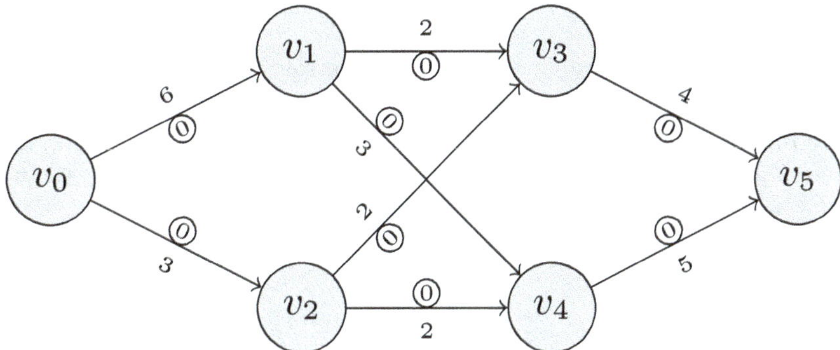

Consider the α-path $v_0v_2v_4v_5$. Apply Activity 7.2.6 **(d)** to this path. What is q?

(b) Adjust the flow on edges by q appropriately. Explain why we need not consider the edge v_2v_4 for any future α-paths.

(c) Pick another α-path where $q > 0$ and repeat until we achieve a maximum flow.

Hint.

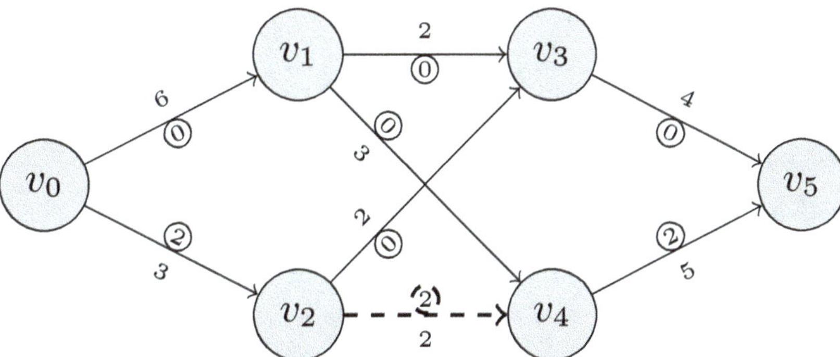

(d) Use the maximum flow and the argument in Activity 7.2.6 to find a minimum cut.

Definition 7.2.10 Ford-Fulkerson Algorithm. We describe an algorithm to find the maximum flow for N, a capacitated network with a unique source v_s and sink v_d:

1. INITIALIZE: We begin with any feasible flow f (including the zero flow).

2. Pick an α-path P in N from v_s to v_d such that:

   - Each forward edge of P satisfies $x_{ij} < c_{ij}$.
   - Each backwards edge satisfies $x_{ij} > 0$.

   (If no such α-path P exists, GOTO 5.)

3. Compute

$$q := \min \left\{ \min_{(v_i, v_j) \text{ a forward edge}} \{c_{ij} - x'_{ij}\}, \min_{(v_j, v_i) \text{ a backwards edge}} \{x'_{ji}\} \right\}.$$

4. Define a new flow f' as follows: for each forward edge of P, (v_i, v_j), we add $x'_{ij} = q + x_{ij}$. For each backwards edge (v_j, v_i) we subtract $x'_{ji} = x_{ji} - q$.
   Let $f := f'$ and GOTO 2.

5. STOP. The flow is now optimal.

$$\Diamond$$

Activity 7.2.11 Prove that the Ford-Fulkerson Algorithm Definition 7.2.10 terminates at a maximum flow.

Definition 7.2.12 Min Cut Algorithm. We describe an algorithm to find the minimum for N, a capacitated network with a unique source v_s and sink v_d:

1. INITIALIZE: We begin with a maximum flow f' and $V_1 = \{v_s\}$.

2. We add v_j to V_1 if there is a $v_i \in V_1$ such that either:

   - $(v_i, v_j) \in E, x'_{ij} < c_{ij}$.
   - $(v_j, v_i) \in E, x'_{ji} > 0$.

   If there is no such v_i, GOTO 4.

3. GOTO 2.

4. Let $V_2 = V \backslash V_1$.

   STOP (V_1, V_2) form a minimum cut.

$$\Diamond$$

7.3 Weighted Graphs

We now pivot to *weighted* networks and consider some natural optimization problems to pose about them.

Exploration 7.3.1 Dr. Ayad is driving from her home to Fantasi College. The town is connected by a series of one way streets, each labeled with the time it would take to traverse the road.

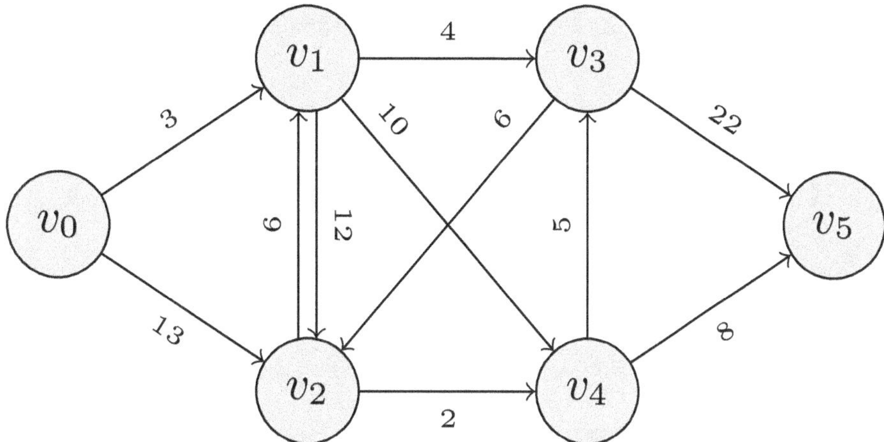

(a) Suppose home for Dr. Ayad is v_0. What is the shortest amount of time needed for her to arrive at Fantasi College on v_5?

(b) Is there a unique route she could take that minimizes this time?

Definition 7.3.2 A network is **weighted** if for each edge $(v_i, v_j) \in E$ we assign (potentially negative!) weight w_{ij}. $\Diamond$

Definition 7.3.3 Give a network N, a **path** P from v_x, v_y is a sequence of consecutive edges $(v_{a_0}, v_{a_1}), \ldots (v_{a_i}, v_{a_{i+1}}), \ldots, (v_{a_{k-1}}, v_{a_k})$ where $v_{a_0} = v_x, v_{a_k} = v_y$. We say that the **length** of P is $\sum_{i=0}^{k-1} w_{a_i a_{i+1}}$. We say that the **distance** from v_x to v_y, $d(v_x, v_y)$, is the length of a shortest path from v_x to v_y. $\Diamond$

Activity 7.3.4 Consider the weighted network:

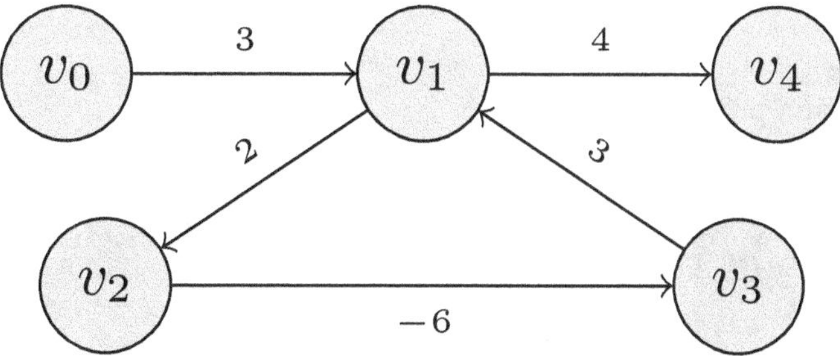

(a) What is the shortest path from v_0 to v_4? (You may repeat edges.)

(b) What if we change w_{23} to -1?

(c) What is a reasonable condition for the shortest path to be well defined?

Definition 7.3.5 We define a **cycle** in a weighted network to be a path from a vertex v_x to itself. If the length of a cycle is negative, we call it a **negative**

cycle. ◇

Activity 7.3.6 In this activity, we model the shortest path problem as a linear optimization problem. Assume N is a weighted network with no negative cycles. Let $0 \leq x_{ij} \leq 1$ where $x_{ij} = 1$ if (v_i, v_j) is in a shortest path P from v_s to v_d.

(a) What is the objective problem?

A. Maximize $\sum_i w_{it} x_{it}$.

B. Maximize $\sum_j w_{sj} x_{sj}$.

C. Maximize $\sum_{i,j} w_{ij} x_{ij}$.

D. Minimize $\sum_i w_{it} x_{it}$.

E. Minimize $\sum_j w_{sj} x_{sj}$.

F. Minimize $\sum_{i,j} w_{ij} x_{ij}$.

(b) What inequality ensures that exactly one edge of the chosen edges is incident to v_d?

A. $\sum_i x_{id} \leq 1$.

B. $\sum_i x_{id} \geq 1$.

C. $\sum_i x_{id} = 1$.

D. $\sum_i x_{id} \leq 0$.

E. $\sum_i x_{id} \geq 0$.

F. $\sum_i x_{id} = 0$.

(c) What inequality ensures that the chosen edges form a path?

A. For each vertex $v_i \neq v_s, v_d$, $\sum_j x_{ji} - \sum_j x_{ij} = 1$.

B. For each vertex $v_i \neq v_s, v_d$, $\sum_j x_{ji} - \sum_j x_{ij} \leq 1$.

C. For each vertex $v_i \neq v_s, v_d$, $\sum_j x_{ji} - \sum_j x_{ij} \geq 1$.

D. For each vertex $v_i \neq v_s, v_d$, $\sum_j x_{ji} - \sum_j x_{ij} = 0$.

E. For each vertex $v_i \neq v_s, v_d$, $\sum_j x_{ji} - \sum_j x_{ij} \leq 0$.

F. For each vertex $v_i \neq v_s, v_d$, $\sum_j x_{ji} - \sum_j x_{ij} \geq 0$.

(d) Why do we not need a constraint for v_s?

(e) Model the shortest path problem in Exploration 7.3.1 as a linear optimization problem and solve it:

```
%display typeset
A = (FIXME)
b = (FIXME)
c = (FIXME)
P = InteractiveLPProblem(A, b, c,
  [FIXME],
  constraint_type = [FIXME],
  variable_type = [FIXME],
  problem_type = FIXME)
P
```

```
print(P.optimal_solution())
print(P.optimal_value())
```

As was the case in previous examples, we introduce a less cumbersome method for finding these shortest paths.

Definition 7.3.7 Dijkstra's Shortest Path Algorithm. Let N be a weighted network with only nonnegative weights. Then *Dijkstra's Shortest Path Algorithm* is as follows:

1. INITIALIZE: Let $R = \{v_s\}$ and let $T = V \backslash R$. Label $\ell_s = 0$, $\ell_i = w_{si}$ if w_{si} exists, ∞ otherwise.

2. Let v_k be a vertex where $\ell_k = \min_{i \in T} \ell_i$.

3. Let $R = R \cup \{v_k\}, T = V \backslash R$.

4. If $T = \emptyset$: STOP.

5. For each $v_j \in T$, let $\ell_j = \min\{\ell_j, \ell_j + w_{kj}\}$.

6. GOTO 2.

When the algorithm terminates, each ℓ_j takes on the value $d(v_s, v_j)$, the length of the shortest path from v_s to v_j. $\qquad\qquad\qquad\qquad\qquad\qquad \Diamond$

Activity 7.3.8 We revisit the shortest path problem from Exploration 7.3.1.

(a) Apply Definition 7.3.7 to the network in this problem and label each vertex v_i by ℓ_i.

(b) What does each ℓ_i represent in terms of travel time?

(c) Consider ℓ_3, ℓ_4 and w_{35}, w_{45}. Which vertex could be on a shortest path from v_0 to v_5?

(d) Take your previous choice of vertex v_k and repeat: look at the ℓ_i of its potential predecessors and w_{ik}. Recursively repeat until we reach v_0.

(e) What is the shortest path from v_0 to v_5?

Activity 7.3.9 We prove that in Definition 7.3.7, $\ell_i = d(v_s, v_i)$ for each $v_i \in R$ via induction on $|R|$.

(a) Verify that the statement is true when $|R| = 1$.

(b) Prove that in step 3, if we select v_k then v_k is adjacent to a vertex in $v_k' \in R$.

(c) Let $|R| = m \geq 1$ and consider v_k, ℓ_k as chosen in step 3. Show that ℓ_k is the shortest distance from v_s to v_k traversing only vertices in $R \cup \{v_k\}$.

(d) Suppose (by way of contradiction) that there was a shortest path P from v_s to v_k where the length of $P < \ell_k$. Show that there must be an edge in P, (v_x, v_y) so that $v_x \in R, v_y \notin R, v_y \neq v_k$.

(e) Show that in this case that $\ell_x + w_{xy} \leq \ell_k$. (Invoke the induction hypothesis.)

(f) Show that the last statement produces a contradiction (why wasn't y already chosen?)

(g) Conclude that $d(v_s, v_k) = \ell_k$.

We present an alternative algorithm for when weights can be negative.

Definition 7.3.10 Shortest Path Algorithm. Let N be a weighted network with no negative cycles. Then an algorithm to find shortest paths is as follows:

1. INITIALIZE: Let $R = \{v_s\}$ and let $T = V \setminus R$. Label $\ell_s = 0$, $\ell_i = w_{si}$ if w_{si} exists, ∞ otherwise.

2. Let v_k be a vertex where $\ell_k = \min\limits_{i \in T} \ell_i$.

3. Let $R = R \cup \{v_k\}, T = V \setminus R$.

4. If $T = \emptyset$: STOP.

5. For each $v_j \in V$, let $\ell_j = \min\{\ell_j, \ell_j + w_{kj}\}$, if $v_j \in R$ has a value changed by this process, remove v_j from R and add it to T.

6. GOTO 2.

When the algorithm terminates, $\ell_j = d(v_s, v_j)$, the length of the shortest path from v_s to v_j. $\diamond$

Exploration 7.3.11 Suppose a shipping company is moving goods through a series of transportation hubs via rail. The maximum capacity in tons and the cost in thousands of dollars per ton are listed as an ordered pair:

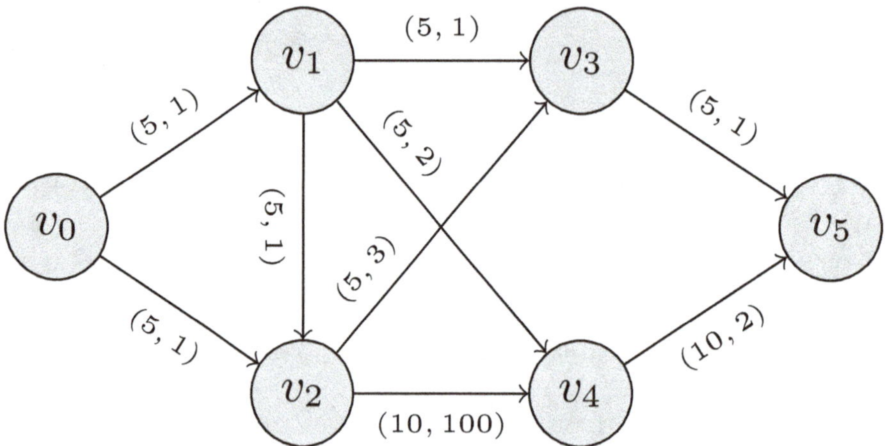

The pairs are (capacity, cost) pairs (denoted (c_{ij}, w_{ij})), and we are trying to ship 10 tons of goods from v_0 to v_5.

(a) Let's find a single path from v_0 to v_5 along which we could ship goods at the lowest possible cost. What criteria should we use to identify this path?

    A. Maximize w_{ij} along this path.   C. Maximize c_{ij} along this path.

    B. Minimize w_{ij} along this path.   D. Minimize c_{ij} along this path.

(b) Find a shortest path from v_0 to v_5.

(c) Use this as an α-path as in Definition 7.2.10.

(d) Decrease capacities of any used edges by x_{ij}.

(e) Repeat (a)-(c) until we have a flow of 10.

(f) Argue that this is *not* the lowest cost flow.

(g) Which of the following was an issue with how this problem was approached?

    A. The original path chosen was too expensive.

    B. The original path forced us into poor choices of future paths.

    C. There was no mechanism to backtrack or adjust previous choices.

Activity 7.3.12 We model the shipping problem in Exploration 7.3.11 as a linear optimization problem. Let x_{ij} denote the quantity in tons of goods shipped from v_i to v_j.

(a) What is the objective function of this problem?

    A. $f(\mathbf{x}) = \sum_{i,j} x_{ij}.$ D. $f(\mathbf{x}) = \sum_{j} x_{0j}.$

    B. $f(\mathbf{x}) = \sum_{i,j} x_{ij} c_{ij}.$ E. $f(\mathbf{x}) = \sum_{j} x_{0j} c_{0j}.$

    C. $f(\mathbf{x}) = \sum_{i,j} x_{ij} w_{ij}.$ F. $f(\mathbf{x}) = \sum_{j} x_{0j} w_{0j}.$

(b) For each ij edge, there is a constraint for the shipping capacity of that edge. What (in)equality represents that capacity?

    A. $x_{ij} \leq w_{ij}.$ D. $x_{ij} \leq c_{ij}.$

    B. $x_{ij} \geq w_{ij}.$ E. $x_{ij} \geq c_{ij}.$

    C. $x_{ij} = w_{ij}.$ F. $x_{ij} = c_{ij}.$

(c) For each vertex v_k excluding the source and sink, what (in)equality represents the conservation of flow?

    A. $\sum_{i} x_{ik} - \sum_{j} x_{kj} \leq 0.$ C. $\sum_{i} x_{ik} - \sum_{j} x_{kj} = 0.$

    B. $\sum_{i} x_{ik} - \sum_{j} x_{kj} \geq 0.$

(d) Let F be the total amount of goods to be shipped, in this case $F = 10$. What equality represents this constraint?

    A. $\sum_{i,j} x_{ij} c_{ij} = F.$ D. $\sum_{j} x_{0j} = F.$

    B. $\sum_{i,j} x_{ij} w_{ij} = F.$ E. $\sum_{j} x_{0j} c_{0j} = 10.$

    C. $\sum_{i,j} x_{ij} = F.$ F. $\sum_{j} x_{0j} w_{0j} = 10.$

(e) Write out the linear optimization problem modeling Exploration 7.3.11 and solve it.

```
%display typeset
A = (FIXME)
b = (FIXME)
c = (FIXME)
P = InteractiveLPProblem(A, b, c,
 [FIXME],
 constraint_type = [FIXME],
 variable_type = [FIXME],
 problem_type = FIXME)
P
```

```
print(P.optimal_solution())
print(P.optimal_value())
```

Activity 7.3.13 We return to Exploration 7.3.11 with an adjustment to the procedure there to enable adjusting previously chosen paths.

(a) Once again, find the shortest path from v_0 to v_5, and use this as an α-path as in Definition 7.2.10.

(b) Now in addition to decreasing the capacities of used edges by x_{ij}, add a backwards edge (v_j, v_i) with capacity x_{ij} and negative weight $-w_{ij}$.

   Pick any path from v_0 to v_5 that traverses a backwards negative edge. What does shipping along this path represent in terms of determining a new shipping procedure.

   Test your ideas out on a few different paths traversing negative edges.

(c) Find the shortest path along this adjusted graph and treat it as an α-path.

(d) What does this second chosen shortest path represent in terms of shipping goods?

(e) Have you now achieved a minimal cost flow shipping 10 tons of goods?

   We now formalize this idea of adjusting previous choices to an algorithm:

Definition 7.3.14 Minimum Cost Flow Algorithm. The steps for the *Minimum Cost Flow Algorithm* are as follows:

1. INITIALIZE: Let $N = (V, E)$ be a weighted capacitated network with a unique source v_s and sink v_d, with no edges going into the source and no edges coming out of the sink. We start with the zero flow $x_{ij} = 0$ for each edge (v_i, v_j). Let F be the desired total flow.

2. If $\varphi(v_d) = \sum_i x_{id} = F$, STOP, we have reached a total flow of F.

3. Form a weighted network $N' = (V', E')$ as follows:

   - Let $V' = V$.
   - Let $(v_i, v_j) \in E'$ if and only if $x_{ij} < c_{ij}$. Let $w'_{ij} = w_{ij}$.
   - Let $(v_j, v_i) \in E'$ if and only if $x_{ij} > 0$. Let $w'_{ij} = -w_{ij}$.

4. Apply Dijkstra's Shortest Path Algorithm on N' to find the shortest path from v_s to v_d. If no path exists STOP, there is no flow with total value F.

5. Find the α-path corresponding to the shortest path found in (4). Let

$$q = \min_{(v_i,v_j)\in\alpha}\left\{\min_{(v_i,v_j)\text{ forward}}\{c_{ij} - x_{ij}\}, \min_{(v_i,v_j)\text{ backwards}}\{F - \varphi(v_d)\}\right\}.$$

6. Add q to each forward x_{ij} in α, and subtract q from each backwards x_{ij} in α.

7. GOTO 2.

$\Diamond$

7.4 Summary of Chapter 7

We use introduce the notion of **networks**, a pair $N = (V, E)$ where V is a set of **vertices** and E is a set of ordered pairs of vertices called **edges**. We also discuss **capacitated networks** where each edge (v_i, v_j) has a **capacity** $c_{ij} \geq 0$. For capacitated networks with designated **sources** and **sinks**, we can define a **flow**, an assignment $0 \leq x_{ij} \leq c_{ij}$ to each edge (v_i, v_j) so that for any non source/sink vertex v_i we have that

$$\sum_j x_{ji} - \sum_j x_{ij} = 0.$$

We focus on capacitated networks with a unique source and sink, with no edges going into the source or out of the sink.

Finding a **maximum flow** on such a network may be solved as a linear optimization problem:

$$\textbf{Maximize: } \sum_i x_{id}$$
$$\textbf{subject to: } \sum_{v_i \in V} x_{ij} - \sum_{v_i \in V} x_{ji} = 0, \text{ for each non source/sink vertex } v_j$$
$$x_{ij} \leq c_{ij}, \text{ for each edge}(v_i, v_j)$$
$$x_{ij} \geq 0, \text{ for each edge}(v_i, v_j).$$

Another problem on such a network is the **minimum cut**, a **cut** is a pair (V_1, V_2) so the V is the disjoint union on V_1, V_2, $v_s \in V_1$ and $v_d \in V_2$. The **capacity** of a cut is the sum

$$\sum_{v_i \in V_1, v_j \in V_2} c_{ij}.$$

Careful analysis of the dual problem to the maximum flow problem shows that the capacity of any cut is an upper bound for the value of any flow.

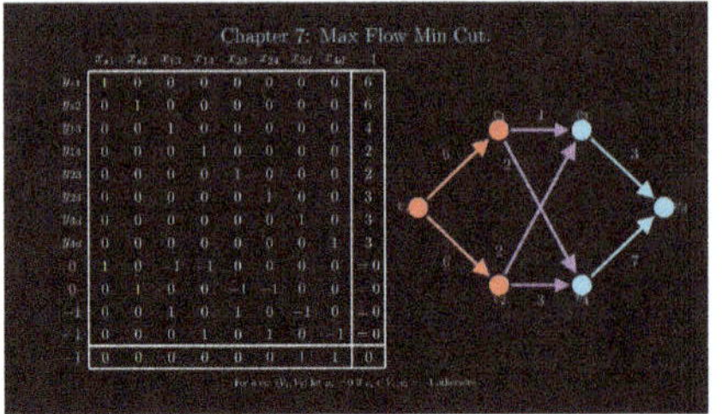

Figure 7.4.1 A linear optimization formulation of maximum flow and minimum cut.

The **Ford-Fulkerson Algorithm** Definition 7.2.10 identifies a maximum flow. Then, starting with $V_1 = \{v_s\}$, we recursively add vertices v_i to V_1 if there is an edge from V_1 to v_i which is not at maximum capacity, or a backwards edge from v_i to V_1. When this is done, we let $V_2 = V \backslash V_1$ and this forms a cut who by construction, has the same capacity as the maximum flow. So by the Weak Duality Theorem, both are optimal.

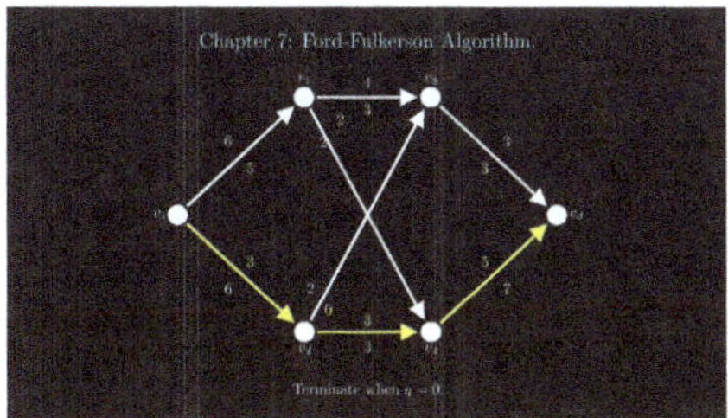

Figure 7.4.2 The Ford-Fulkerson algorithm and finding max flows/min cuts.

Another type of network is a **weighted** network, where each edge (v_i, v_j) has a potentially negative weight w_{ij}. **Dijkstra's Shortest Path Algorithm** Definition 7.3.7 describes an algorithm which identifies the distance (sum of weights) from a starting source v_s to any other vertex v_i in the network, by labeling each vertex with the current shortest distance from v_s to v_i and relabeling and readjusting as shorter distances are found.

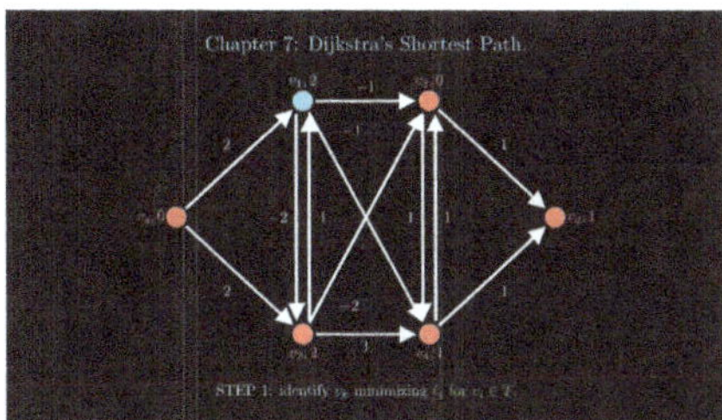

Figure 7.4.3 Dijkstra's Shortest Path algorithm and finding the shortest path between vertices.

The shortest path algorithm has a useful application in the Minimum Cost Flow Algorithm Definition 7.3.14. In this problem, we try to find a flow of value F from v_s to v_d on a weighted capacitated network that minimizes the cost to do so. One can identify shortest paths from v_s to v_d and increase flows along this path, and repeat recursively. However, doing so greedily may result in nonoptimal solutions. We construct a second network N' where backwards edges with negative weight are added for flows on N to represent the ability to reduce flow along edges, and the shortest paths are found on N'. Doing so repeatedly results in the actual minimum cost flow.

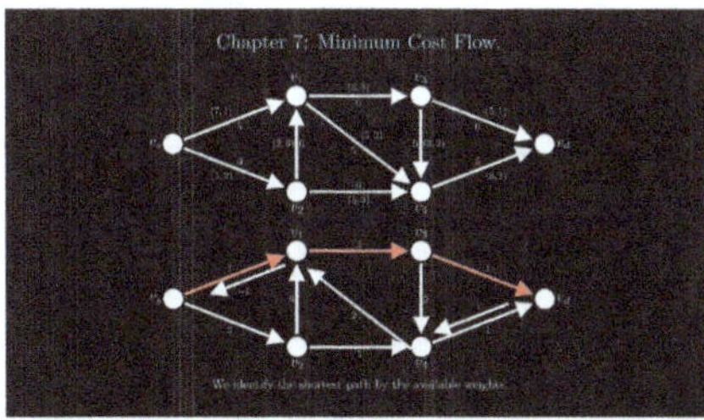

Figure 7.4.4 Minimum Cost Flow algorithm and finding the minimum cost flow of a given value.

7.5 Problems for Chapter 7

Exercises

1. Consider the following capacitated network and given flow:

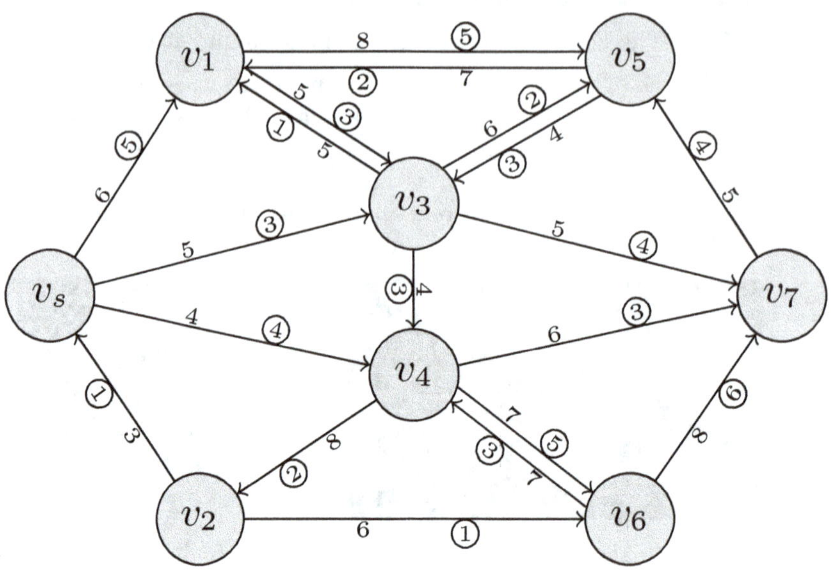

(a) Find $\varphi(v_i)$ for each vertex, and compute $\displaystyle\sum_{i=1}^{7} \varphi(v_i)$.

(b) Classify each vertex as a source, sink or intermediate vertex.

(c) Add two vertices and as few edges as possible to extend this flow to a flow with a unique source and unique sink

2. For each of the following capacitated networks, find the max-flow and min-cut on these networks as shown in Section 7.2.

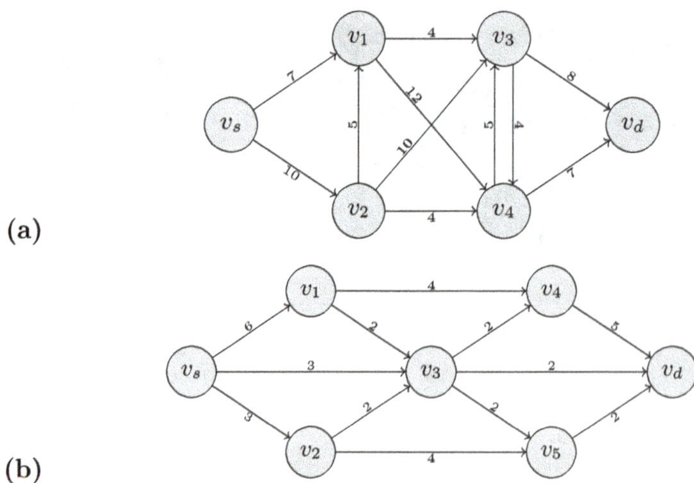

(a)

(b)

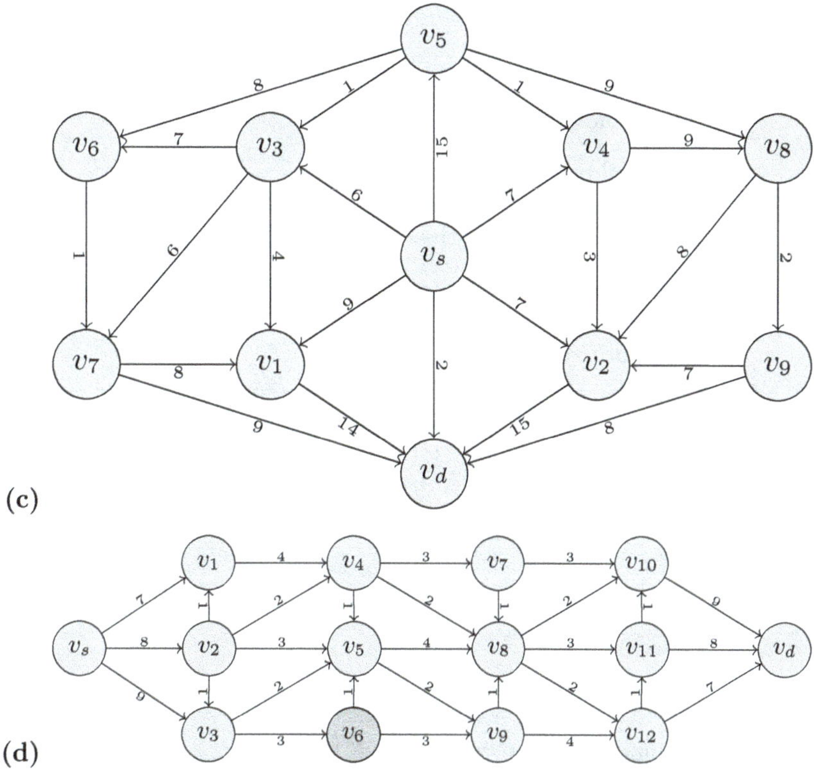

(c)

(d)

3. Consider the max-flow min-cut problem Exercise 7.5.2 **(a)**.

   (a) Write out the noncanonical maximization problem which would compute the max-flow.

   (b) Record this problem in a Tucker tableau, then record the dual variable using $\boxed{\mu_i}$ to denote dual variables associated with vertex equality constraints and y_{ij} to denote the dual variables for edge inequality constraint.

   (c) Verify that the max-flow and min-cut you found are feasible solutions to these problems (using the convention that $\mu_k = 0$ if $v_k \in V_1$, $\mu_k = -1$ if $v_k \in V_2$ and $y_{ij} = 1$ if $v_i \in V_1, v_j \in V_2$ and $y_{ij} = 0$ otherwise.)

   (d) Argue that any cut corresponds to a feasible solution to this dual problem.

   (e) How can we tell both the flow and cut found in Exercise 7.5.2 **(a)** are optimal? (Think duality.)

4. Over a month at a hospital many patients are in need of blood transfusions. They had available blood from 47 donors with type A blood, 33 donors with type B blood, 46 donors with type AB and 44 donors with type O. There were 38 patients with type A blood, 39 patients with type B blood, 49 patients with type AB and 43 patients with type O. Type A patients can only receive type A or O, type B patients can receive only type B or O, type O patients can receive only type O, and type AB patients can receive any of the four types.

   (a) Construct a capacitated network which models the distributions of blood type from donors to patients with a unique source and sink, with no edges into the source or out of the sink.

(b) Find a max-flow for this network.

(c) Find a min-cut for this network.

(d) If not all patients were able to receive blood, explain to the financial backers and hospital administrators, who may not have taken any math in awhile, why this was the case.

(e) There's little a hospital can do about the blood types of incoming patients. If reaching out to potential donors, what blood types should be prioritized?

5. Let N be a capacitated network with a unique source and sink, with no edges going into the source or out of the sink. Let x_{ij} be a flow on this network with value f, and (V_1, V_2) be a cut of this network. Then prove that

$$f = \sum_{v_i \in V_1, v_j \in V_2} x_{ij} - \sum_{v_i \in V_1, v_j \in V_2} x_{ji}.$$

Hint. What is the sum $\sum_{v_j \in V_2} \varphi(v_j)$? How can rewriting this as a double sum help?

6. Let N be a capacitated network with a unique source and sink, with no edges going into the source or out of the sink.

(a) Given an example for N such that N has nonunique max-flows.

(b) Given an example for N such that N has nonunique min-cuts.

(c) Let x_{ij} denote any max-flow for N with value f and (V_1, V_2) denote any min-cut (not necessarily produced by x_{ij} and Definition 7.2.12). Let (V_1', V_2') be the cut generated x_{ij} via Definition 7.2.12.

Prove that $x_{ij} = c_{ij}$ for any $v_i \in V_1, v_j = V_2$.

Hint. Use Exercise 7.5.5.

(d) Let x_{ij}, x_{ij}' be two distinct max-flows on N, and $(V_1, V_2), (V_1', V_2')$ be the cuts produced by Definition 7.2.12 on with these flows respectively. Prove that $(V_1, V_2) = (V_1', V_2')$.

Hint. Use previous part.

7. For each of the following, find the shortest path from v_s to v_d using Definition 7.3.7.

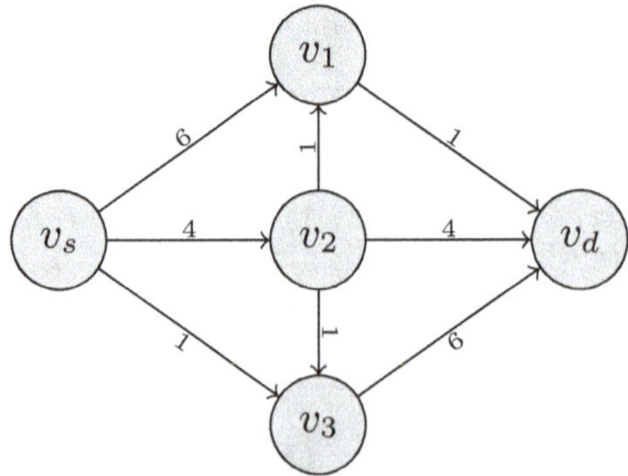

(a)

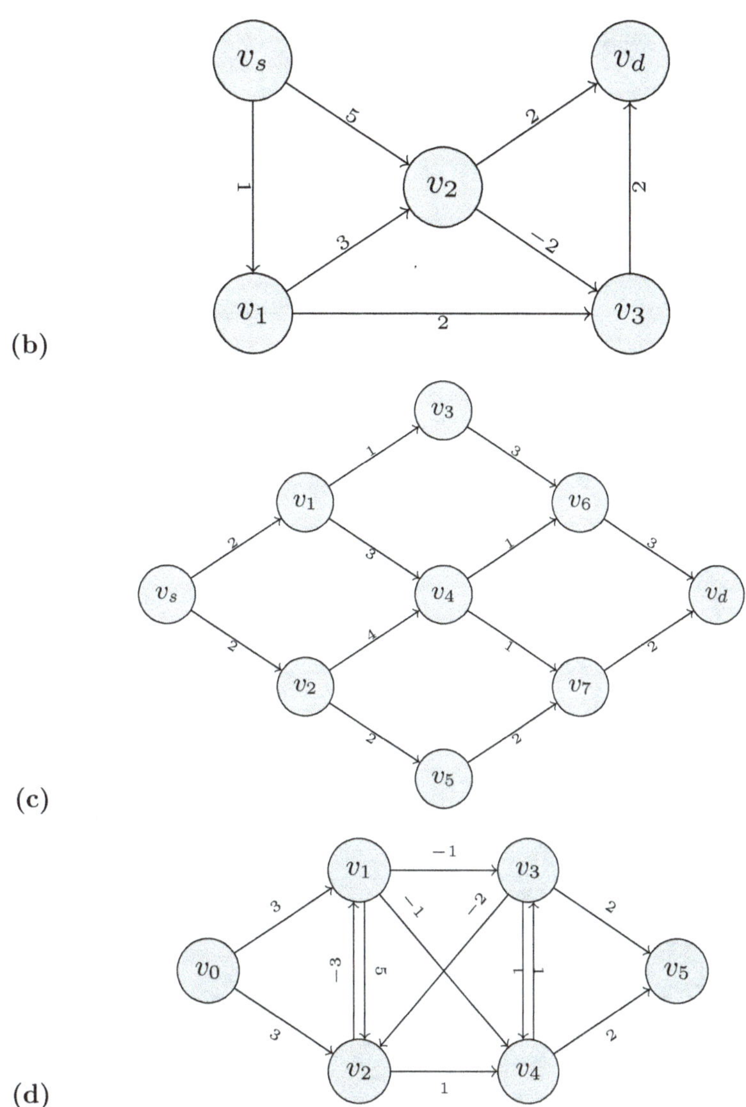

(b)

(c)

(d)

8. Consider Exercise 7.5.7 **(a)**. Model this problem as a linear optimization problem and solve.

9. Let N be a weighted network with positive weights. For the following, prove or find a counterexample:

   (a) Let $x, y, z \in V$. Prove that the value of the shortest path from x to z is the sum of the value of the shortest path from x to y plus the value of the shortest path from y to z.

   (b) Suppose there was an edge going into v_s, then Definition 7.3.7 fails.

10. For each of the following, find the minimum cost-flows for $F = 8$ and $F = 10$. Interpret each ordered pair (c_{ij}, w_{ij}) as (capacity,cost).

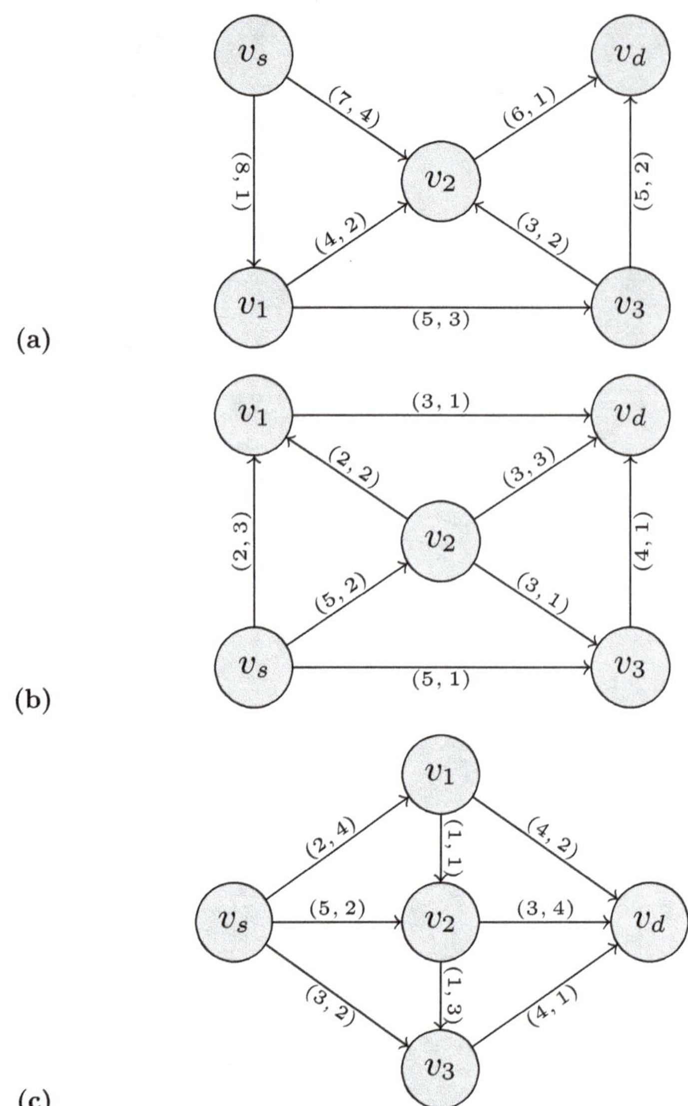

11. Model Exercise 7.5.10 **(a)** as a linear optimization problem.

12. For each problem in Exercise 6.5.1, draw a weighted capacitated network where the transportation problem may be solved by solving an appropriate min-cost flow problem. State what the value F of the flow should be. Do not solve the problem.

13. For each problem in Exercise 6.5.9, draw a weighted capacitated network where the transportation problem may be solved by solving an appropriate min-cost flow problem. State what the value F of the flow should be. Do not solve the problem.

Chapter 8

Integer Optimization

Thus far, we have been exploring problems where the solution space consists of real-valued vectors. However, in some contexts, like in Chapter 6 and Chapter 7, it is sensible to imagine that fractional or irrational solutions may not make sense in the real-world contexts of those problems. It's also not a stretch to imagine more classical production-type problems where only integral units would be possible.

In the cases where real-valued and integer-valued optimal solutions may differ, some care must be taken to solve for the optimal integer valued solution. In Section 8.1, we explore one potential algorithm which is algebraically driven, and in Section 8.2, we discuss an alternative algorithm which is geometrically focused.

8.1 Branch and Bound Method

We begin motivating integer optimization problem, and explore an algebraically centered method for solving them.

Exploration 8.1.1 Suppose the witch Agnesi also goes into the business of selling food: meat sandwiches and meat pies. Each day she is able to acquire 50 oz of mystery meat (don't ask) and 30 oz of flour. The sandwiches take 8 oz of meat and 2 oz of flour, the pies take 3oz of meat and 5 oz of flour. She is able to sell the sandwiches for 10 gold pieces and the pies for 7 gold pieces.

(a) Agnesi wishes to be able to produce sandwiches and pies in a way to maximize her income. Set up this problem as a linear optimization problem and solve.

```
%display typeset
A = (FIXME)
b = (FIXME)
c = (FIXME)
P = InteractiveLPProblem(A, b, c,
  [FIXME],
  constraint_type = [FIXME],
  variable_type = [FIXME],
  problem_type = FIXME)
P
```

```
print(P.optimal_solution())
print(P.optimal_value())
```

(b) What are some problems with this solution? Without trying to explicitly find the optimal solution, estimate how far off your current solution is from the optimal.

(c) How many sandwiches and pies should she actually sell to maximize her income?

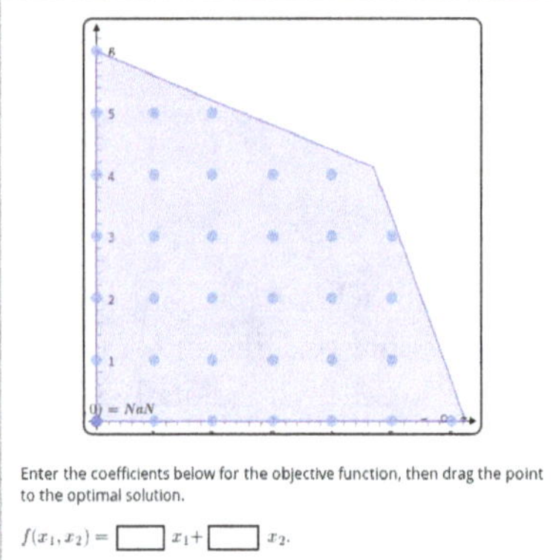

(d) After the local barber has been arrested, demand for Agnesi's pies sees an increase, and she is able to now sell them for 12 gold pieces. Now what production level maximizes her income?

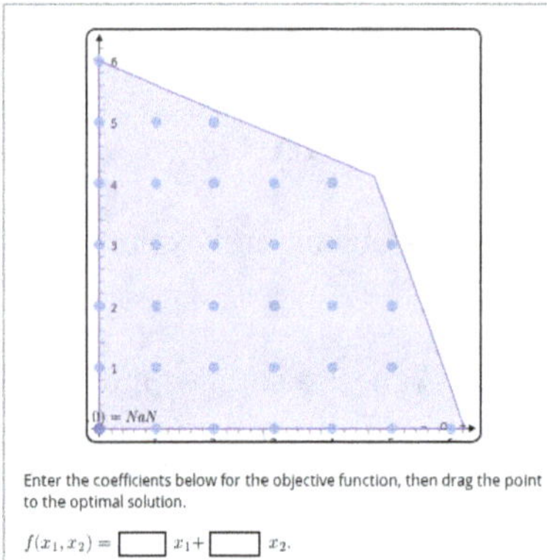

Definition 8.1.2 A linear optimization problem where all solutions must be integers is called an **integer optimization problem**. If the condition that solutions be integers are relaxed, this is called the **relaxation** of the integer optimization problem. ◊

Activity 8.1.3

(a) Come up with three realistic optimization problems which are best modeled by an integer optimization problem. You do not need to work out all the details or solve the problems.

(b) Consider an integer optimization maximization problem whose relaxation achieves an optimal solution. Which of the following are true about the integer optimization problem?

    A. The integer problem achieves an optimal solution that is equal to the optimal solution of the relaxation.

B. The integer problem achieves an optimal solution that is less than to the optimal solution of the relaxation.

C. The integer problem achieves an optimal solution that is greater than to the optimal solution of the relaxation.

D. The integer problem is infeasible.

E. The integer problem is unbounded above.

Activity 8.1.4 The **branch and bound** method is a way to systematically add bounds to a linear optimization problem to ensure the solution is integral.

We examine the integer optimization problem from Exploration 8.1.1, and its relaxation.

(a) Let x_1 be the number of sandwiches sold, and consider the current optimal x_1 from the relaxation. Which two of the following potential additional constraints would force the value of x_1 to be an integer, while remaining as close to the optimal solution of the relaxation as possible.

A. $x_1 \leq 3$.	E. $x_1 \leq 5$.
B. $x_1 \geq 3$.	F. $x_1 \geq 5$.
C. $x_1 \leq 4$.	G. $x_1 \leq 6$.
D. $x_1 \geq 4$.	H. $x_1 \geq 6$.

(b) Let us consider the additional constraint $x_1 \leq 4$. Solve the resulting general linear optimization problem:

```
%display typeset
A = (FIXME)
b = (FIXME)
c = (FIXME)
P = InteractiveLPProblem(A, b, c,
  [FIXME],
  constraint_type = [FIXME],
  variable_type = [FIXME],
  problem_type = FIXME)
P
```

```
print(P.optimal_solution())
print(P.optimal_value())
```

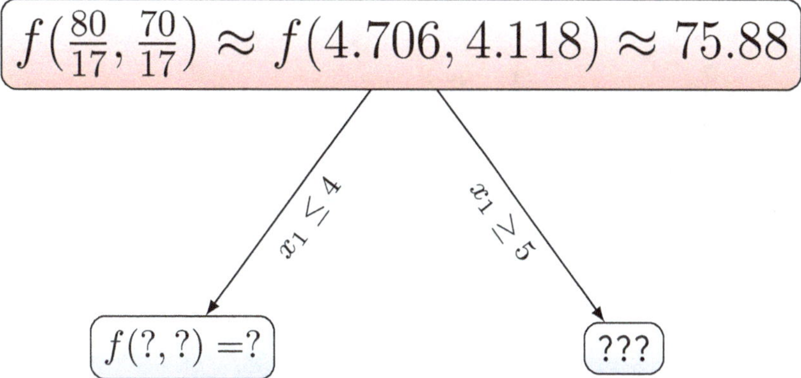

(c) What additional constraints on x_2 would ensure that x_2 would be integral while changing x_2 as little as possible?

(d) Consider the additional constraint $x_2 \leq 4$. Solve this maximization problem. Why do we no longer need to adjust x_1? Are we guaranteed that this solution is optimal?

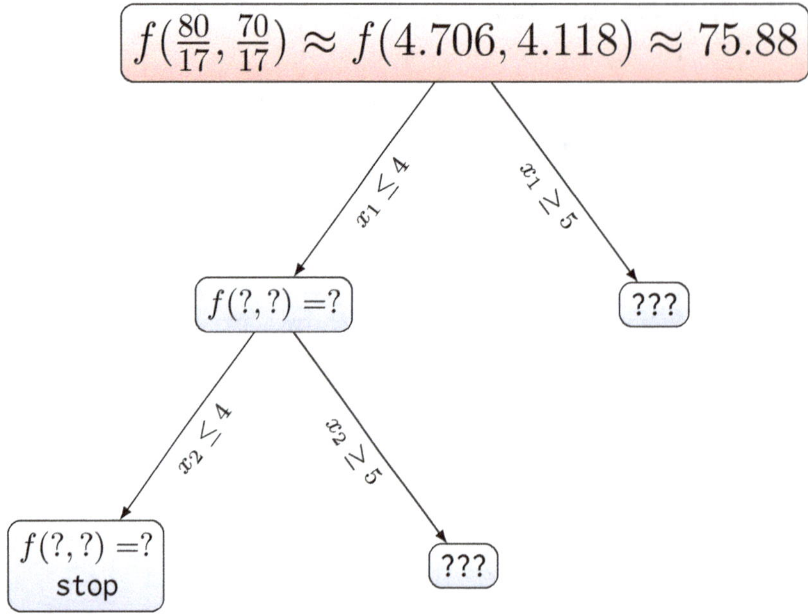

(e) We consider the constraint $x_2 \geq 5$ instead. Solve this maximization problem. Even though the solution is not integral, why do we no longer need to pursue the case where $x_1 \leq 4, x_2 \geq 5$?

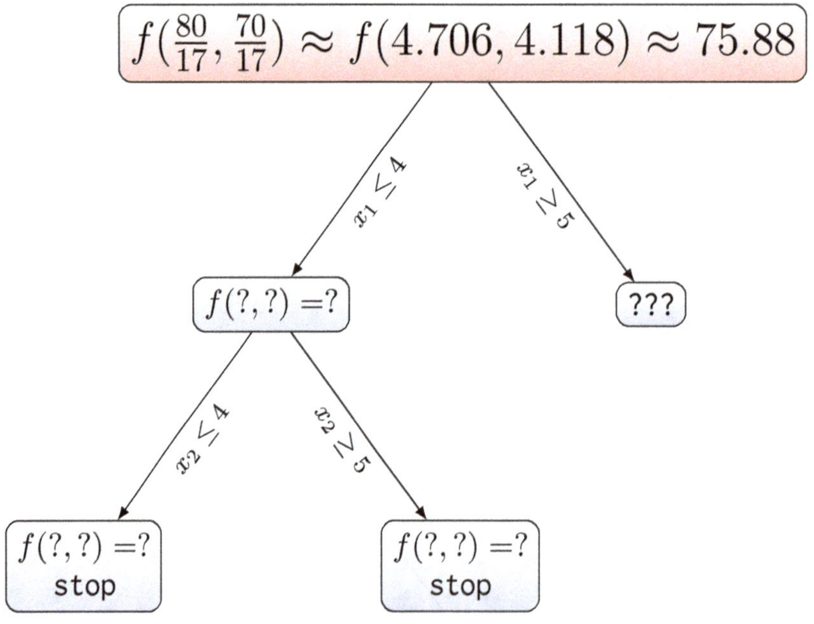

(f) We return to the other possible initial constraint for x_1, $x_1 \geq 5$. Solve this maximization problem. What are the possible constraints we could now add for x_2?

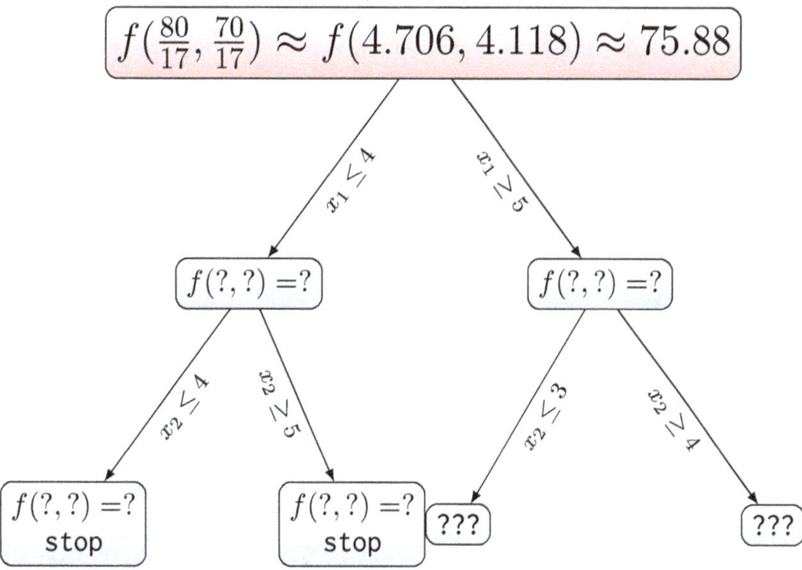

(g) We consider now the constraint $x_2 \leq 3$. Solve this maximization problem. If this solution integral?

(h) Now in addition to the $x_2 \leq 3$, we have two choices for an additional constraint on x_1, one $\leq$ constraint, and one $\geq$ constraint. Solve the system with each of these additional constraints. If any of the solutions are integral, how do they compare with previous integral solutions that we have found?

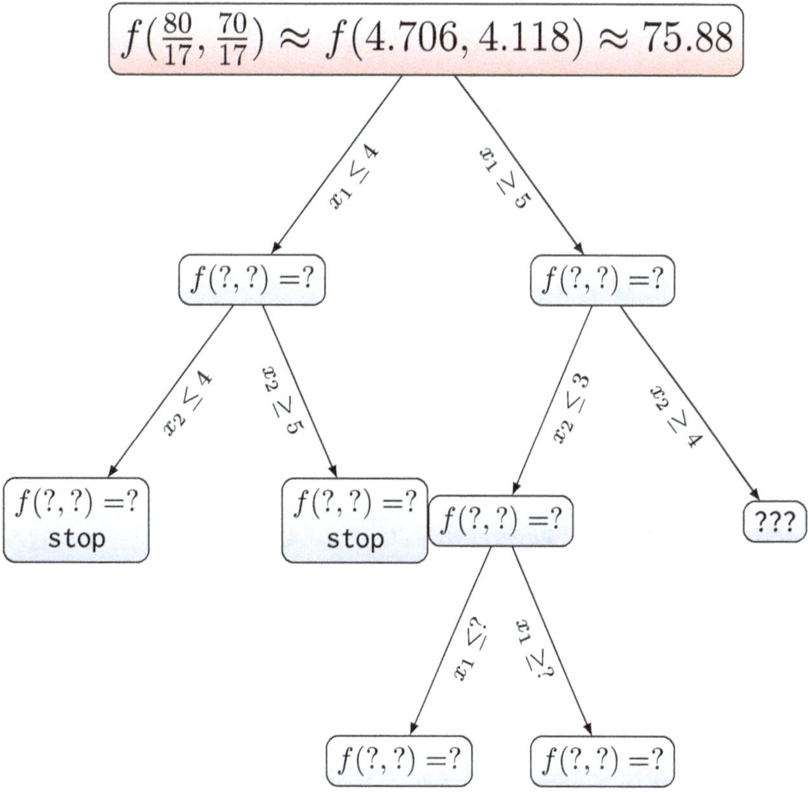

(i) One of the x_1 constraints in the previous task results in an nonintegral solution. Nevertheless, we no longer need to continue to add constraints to this problem to obtain an integral solution. Why not?

(j) Now instead of the $x_2 \leq 3$ constraint we had above, we now consider the constraint $x_2 \geq 4$. Solve this maximization problem. Why do we no longer need to consider the case where $x_2 \geq 4$?

(k) How do we know we now have arrived at the optimal solution?

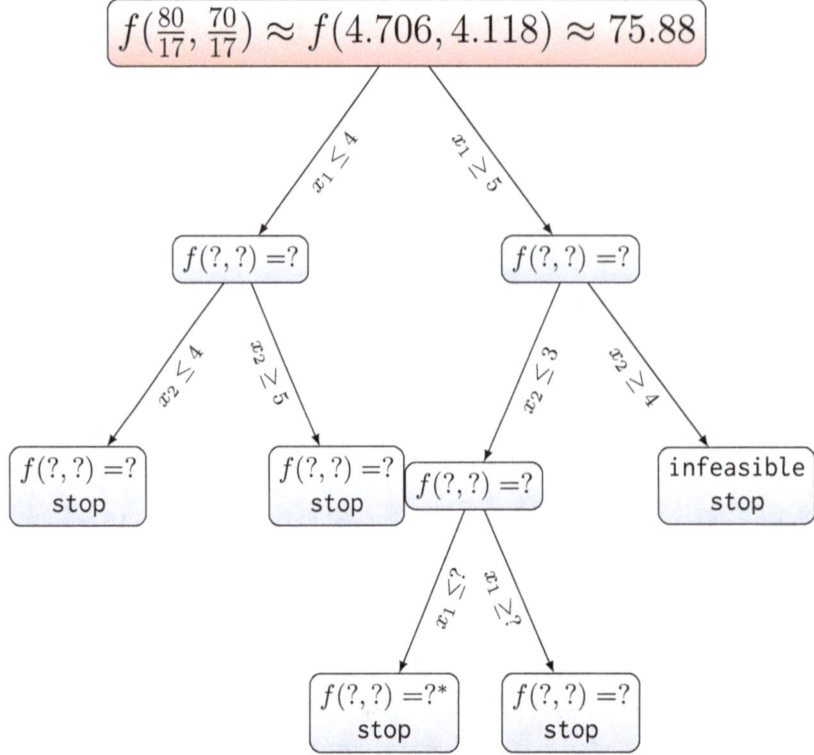

Definition 8.1.5 The **Branch and Bound Algorithm** for solving an integer optimization maximization problem is as follows:

1. INITIALIZE: Begin with a canonical maximization integer optimization problem.

2. Solve the relaxation of the linear optimization problem. If the solution is integral: STOP.

3. For some x_i^* in the optimal solution found in the previous step, define two sub problems, one with additional constraint $x_i \leq \lfloor x_i^* \rfloor$ and $x_i \geq \lceil x_i^* \rceil$.

4. Pick one of the subproblems and solve the linear relaxation with the additional constraint.

5. If the solution is integral, RETURN to 4.

6. If the solution is less than any integral solution found, RETURN to 4.

7. If the problem is infeasible, RETURN to 4.

8. Apply 3-7 to the new problem.

9. If all subproblems are explored, RETURN to 4 for the parent problem.

10. Once the search tree has been exhausted, identify the optimal integral solution.

$\Diamond$

Example 8.1.6 The complete search tree for Activity 8.1.4 is as follows:

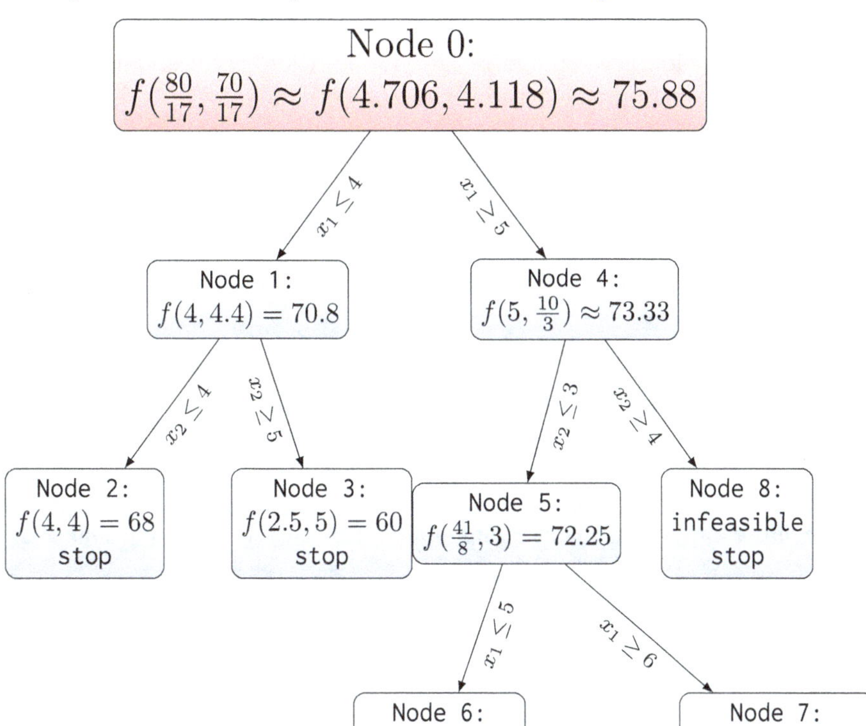

We start at Node 0, and identify the two subproblems. Next, we examine the subproblem where $x_1 \leq 4$ in Node 1, and again identify two subproblems. Examining Node 2 causes us to stop and return because the solution was integral. When examining Node 3, we also stop and return even though the solution is not integral, since the optimal solution for that subproblem was already less than the solution found in Node 1, and any further exploration would lead to a lower value still.

Next, we return to the starting node and to the other initial subproblem in Node 4, were $x \geq 5$. When we split into the two subproblems, $x_2 \leq 3$ and $x_2 \geq 4$. $x_2 \leq 3$gives a nonintegral solution in Node 5, but since it is greater than the integral solution found in Node 2, we continue.

The two additional constraints we may not add to Node 5 are $x_1 \leq 5$ and $x_1 \geq 6$. In Node 6, $x_1 \leq 5$ results in an integral solution greater than the one found in Node 2, and we may return. In Node 7, $x_1 \geq 6$ yields a nonintegral solution less than the integral solution found in Node 6, there is no point in continuing and we return.

Returning back to Node 4, we proceed to Node 8, with the constraint $x_2 \geq 4$, rather than the constraint $x_2 \leq 3$. However this yields an infeasible problem and we may return, as there is no point in adding additional constraints.

Thus, we return, and of the two integral solutions found, $f(5, 3) = 71$ has the highest value.　　　　$\square$

Activity 8.1.7 As the demand for meat pies skyrockets, Agnesi is now able to acquire 40 oz of floor a day, but now uses 10 oz of floor per meat pie to thicken the gravy. She is able to sell these for 40 gp each. Use the Branch and Bound Algorithm Definition 8.1.5 to find her new optimal production level.

8.2 Cutting-Plane Method

We continue our journey through integer optimization, and examine a second method to solve these problems which is geometrically oriented.

Exploration 8.2.1 Recall Exploration 8.1.1, and the question of making sandwiches and pies.

Define two additional inequalities such that the following are true:

A. No inequality eliminates any feasible integer solution of the original problem.

B. No boundary hyperplane is parallel to the objective function plane.

C. With the additional inequalities, the optimal solution to the linear relaxation is the optimal integer solution previously found for Exploration 8.1.1.

The boundary for these additional inequalities are referred to as **cutting hyperplanes**. We wish to determine how to find such cutting hyperplanes.

Activity 8.2.2 In this activity, we motivate the math behind the cutting-plane method.

Let $x_1^*, \ldots, x_m^*, \ldots, x_{m+n}^*$ be a feasible solution of the relaxation of a canonical integer maximization problem, where the x_i are basic (slack) variables and the x_j are nonbasic variables.

We consider the constraint

$$x_i + \sum_j a_{ij} x_j = b_i.$$

	x_j		-1	
$\cdots$	$\cdots$	$\cdots$	$\vdots$	$= -?$
$\cdots$	a_{ij}	$\cdots$	b_i	$= -x_i$
$\cdots$	$\cdots$	$\cdots$	$\vdots$	$= -?$
$\cdots$	$\cdots$	$\cdots$	$?$	$= f$

(a) Explain why the above equality is equivalent to

$$x_i + \sum_j \lfloor a_{ij} \rfloor x_j - \lfloor b_i \rfloor = (b_i - \lfloor b_i \rfloor) - \sum_j (a_{ij} - \lfloor a_{ij} \rfloor) x_j$$

(b) Show that for any feasible integral solution, that the left hand side of the equality in **(a)** is an integer.

(c) Show that the right hand side of the equation in **(a)** is strictly less than 1 for any feasible solution.

(d) For any integral solution, what is a nonnegative integer upper bound for $(b_i - \lfloor b_i \rfloor) - \sum_j (a_{ij} - \lfloor a_{ij} \rfloor) x_j$?

(e) Show that

$$\sum_j (\lfloor a_{ij} \rfloor - a_{ij}) x_j \leq (\lfloor b_i \rfloor - b_i)$$

for any feasible integral solution to the relaxation of the integer optimization problem.

We call the hyperplane $\sum_j (\lfloor a_{ij} \rfloor - a_{ij}) x_j = (\lfloor b_i \rfloor - b_i)$ a **cutting-plane**.

(f) Show that if b_i is nonintegral, then by adding this constraint, the solution $x_1^*, \ldots, x_m^*, \ldots, x_{m+n}^*$ is no longer feasible.

Activity 8.2.3 We now apply this idea to an integer problem. Consider the integer optimization problem:

$$\textbf{Maximize: } f(x_1, x_2) = -x_1 + 5x_2$$
$$\textbf{Subject to: } 3x_1 + 2x_2 \leq 12$$
$$-3x_1 + 2x_2 \leq 7$$
$$x_1, x_2 \geq 0$$
$$x_1, x_2 \text{ are integers.}$$

(a) Solve the relaxation of this integer problem, and verify that this solution is not integral.

t_1	t_2	-1	
a_{11}	a_{12}	b_1	$= -x_1$
a_{21}	a_{22}	b_2	$= -x_2$
c_1	c_2	d	$= f$

(b) Take the second row $x_2 = b_2 - a_{21}t_1 - a_{22}t_2$ and follow the procedure in Activity 8.2.2 to generate a new constraint $a_{31}t_1 + a_{32}t_2 \leq b_3$:

t_1	t_2	-1	
a_{11}	a_{12}	b_1	$= -x_1$
a_{21}	a_{22}	b_2	$= -x_2$
a_{31}	a_{32}	b_3	$= -t_3$
c_1	c_2	d	$= f$

(c) Using the fact that $t_1 = 12 - 3x_1 - 2x_2, t_2 = 7 + 3x_1 - 2x_2$, describe this cutting-plane $a_{31}t_1 + a_{32}t_2 = b_3$ in terms of x_1, x_2.

(d) Pivot on the entry a_{31} and verify that the resulting basic solution is optimal and nonintegral.

t_3	t_2	-1	
a_{11}	a_{12}	b_1	$= -x_1$
a_{21}	a_{22}	b_2	$= -x_2$
a_{31}	a_{32}	b_3	$= -t_1$
c_1	c_2	d	$= f$

(e) There is only one valid choice of row to generate a new constraint. Follow the procedure in Activity 8.2.2 to generate a new constraint $a_{41}t_3 + a_{42}t_2 = b_4$.

$$
\begin{array}{cc|c}
t_3 & t_2 & -1 \\
\hline
a_{11} & a_{12} & b_1 \\
a_{21} & a_{22} & b_2 \\
a_{31} & a_{32} & b_3 \\
a_{41} & a_{42} & b_4 \\
\hline
c_1 & c_2 & d
\end{array}
\begin{array}{l}
\\
= -x_1 \\
= -x_2 \\
= -t_1 \\
= -t_4 \\
= f
\end{array}
$$

(f) Using the fact that $t_3 = -\frac{3}{4} + \frac{1}{4}t_1 + \frac{1}{4}t_2$, express this new cutting-plane $a_{41}t_3 + a_{42}t_2 = b_4$ in terms of x_1, x_2.

(g) Pivot on the entry a_{42} and verify that the resulting basic solution is optimal and nonintegral.

$$
\begin{array}{cc|c}
t_3 & t_4 & -1 \\
\hline
a_{11} & a_{12} & b_1 \\
a_{21} & a_{22} & b_2 \\
a_{31} & a_{32} & b_3 \\
a_{41} & a_{42} & b_4 \\
\hline
c_1 & c_2 & d
\end{array}
\begin{array}{l}
\\
= -x_1 \\
= -x_2 \\
= -t_1 \\
= -t_2 \\
= f
\end{array}
$$

(h) Use either the t_1 or t_2 row to generate a new constraint $a_{51}t_3 + a_{52}t_4 \le b_5$.

$$
\begin{array}{cc|c}
t_3 & t_4 & -1 \\
\hline
a_{11} & a_{12} & b_1 \\
a_{21} & a_{22} & b_2 \\
a_{31} & a_{32} & b_3 \\
a_{41} & a_{42} & b_4 \\
a_{51} & a_{52} & b_5 \\
\hline
c_1 & c_2 & d
\end{array}
\begin{array}{l}
\\
= -x_1 \\
= -x_2 \\
= -t_1 \\
= -t_2 \\
= -t_5 \\
= f
\end{array}
$$

(i) Using the fact that $t_4 = -\frac{1}{3} + \frac{2}{3}t_3 + \frac{2}{3}t_2$, express this new cutting-plane $a_{51}t_3 + a_{52}t_4 = b_5$ in terms of x_1, x_2.

(j) Pivot on the a_{52} entry and verify that the resulting basic solution is optimal *and* integral!

$$
\begin{array}{cc|c}
t_3 & t_5 & -1 \\
\hline
a_{11} & a_{12} & b_1 \\
a_{21} & a_{22} & b_2 \\
a_{31} & a_{32} & b_3 \\
a_{41} & a_{42} & b_4 \\
a_{51} & a_{52} & b_5 \\
\hline
c_1 & c_2 & d
\end{array}
\begin{array}{l}
\\
= -x_1 \\
= -x_2 \\
= -t_1 \\
= -t_2 \\
= -t_3 \\
= f
\end{array}
$$

(k) Enter the coefficients for the objective function and the three cutting-planes in order that you found them, and then drag the point to the optimal solution to the integer problem.

Definition 8.2.4 Gomory Cutting-Plane Algorithm. The **Gomory Cutting-Plane** Algorithm for an integer optimization problem is as follows:

1. INITIALIZE: Begin with a canonical maximization integer optimization problem.

2. Solve the relaxation of the integer problem. If all the resulting b_i are integral STOP: you have found an optimal integral solution.

3. Select a b_i that is nonintegral and for that row, construct the additional bound: $\sum_j (\lfloor a_{ij} \rfloor - a_{ij}) x_j \leq (\lfloor b_i \rfloor - b_i)$.

4. GOTO 2.

$\Diamond$

8.3 Solving Integer Optimization Problems with Sage

In Section 2.4 and Section 3.3, we solved canonical and noncanonical linear optimization problems. Solving such problems by hand could be tedious, and the techniques to solve integer optimization problems are even more involved. In this section, we solve integer optimization problems using Sage.

Since integer optimization is more difficult computationally than linear optimization, we use different commands to find solutions. Rather than use `InteractiveLPProblem`, we use `MixedIntegerLinearProgram`.

Activity 8.3.1 Say we want to solve the integer optimization problem:

$$\textbf{Minimize: } f(\mathbf{x}) = 3x_1 + 4x_2 + 2x_3$$
$$\textbf{subject to: } x_1 \leq 7$$
$$x_2 + x_3 \geq 5$$
$$5x_1 + 3x_2 + 2x_3 \leq 37$$
$$x_1, x_2, x_3 \geq 0, \text{ and are integers.}$$

(a) Record this noncanonical problem using Sage:

```
P = MixedIntegerLinearProgram(solver="GLPK")
X = P.new_variable(integer=True, nonnegative=True)
P.add_constraint(X[1] <= 7)
P.add_constraint(X[2]+X[3] >= 5)
P.add_constraint(5*X[1]+3*X[2]+2*X[3] <= 37)
P.set_objective(3*X[1]+4*X[2]+2*X[3])
P.show()
```

(b) We can then find the optimal solution:

```
print(P.solve())
for i, v in sorted(P.get_values(X, convert=ZZ,
    tolerance=1e-3).items()):
    print(f'x_{i} = {v}')
```

(c) We could also minimize f if we chose:

```
P = MixedIntegerLinearProgram(solver="GLPK",
    maximization=False)
X = P.new_variable(integer=True, nonnegative=True)
P.add_constraint(X[1] <= 7)
P.add_constraint(X[2]+X[3] >= 5)
P.add_constraint(5*X[1]+3*X[2]+2*X[3] <= 37)
P.set_objective(3*X[1]+4*X[2]+2*X[3])
P.show()
```

```
print(P.solve())
for i, v in sorted(P.get_values(X, convert=ZZ,
    tolerance=1e-3).items()):
    print(f'x_{i} = {v}')
```

(d) We could also have solved the linear relaxation of the original problem if we chose:

```
P = MixedIntegerLinearProgram(solver="GLPK")
X = P.new_variable(integer=False, nonnegative=True)
P.add_constraint(X[1] <= 7)
P.add_constraint(X[2]+X[3] >= 5)
P.add_constraint(5*X[1]+3*X[2]+2*X[3] <= 37)
P.set_objective(3*X[1]+4*X[2]+2*X[3])
P.show()
```

```
print(P.solve())
for i, v in sorted(P.get_values(X).items()):
    print(f'x_{i}␣=␣{v}')
```

Activity 8.3.2 Solve:

$$\textbf{Maximize: } f(\mathbf{x}) = 3x_1 + 2x_2$$
$$\textbf{subject to: } 4x_1 + 5x_2 \le 39$$
$$7x_1 + 3x_2 \ge 20$$
$$x_2 \le 5$$
$$x_1, x_2 \text{ are integers.}$$

```
P = MixedIntegerLinearProgram(solver="GLPK")
X = P.new_variable(FIXME)
P.add_constraint(FIXME)
P.set_objective(FIXME)
P.show()
```

```
print(P.solve())
for i, v in sorted(P.get_values(X, convert=ZZ,
    tolerance=1e-3).items()):
    print(f'x_{i}␣=␣{v}')
```

8.4 Summary of Chapter 8

An **integer optimization problem** is an optimization problem where all solutions must only have integer values. We generally begin by solving the **relaxation** of an integer problem, where there is no such restriction. If the optimal solution is integral, then no additional work is needed. Otherwise, there are techniques which allow us to find the integral solutions.

The **Branch and Bound** method Definition 8.1.5 works by recursively adding constraints to the problem. If for some optimal solution $x_i = x_i^*$ is nonintegral, then we may force it to be integral with the additional constraint that $x_i \leq \lfloor x_i^* \rfloor$ or $x_i \geq \lceil x_i^* \rceil$. Each choice results in a new branch of a search tree. If the additional constraint results in an integral optimal solution or an infeasible solution, no additional constraints are needed and we may return to the parent node. We terminate when all branches are explored in this way. Then amongst the integer solutions, we select the optimal choice.

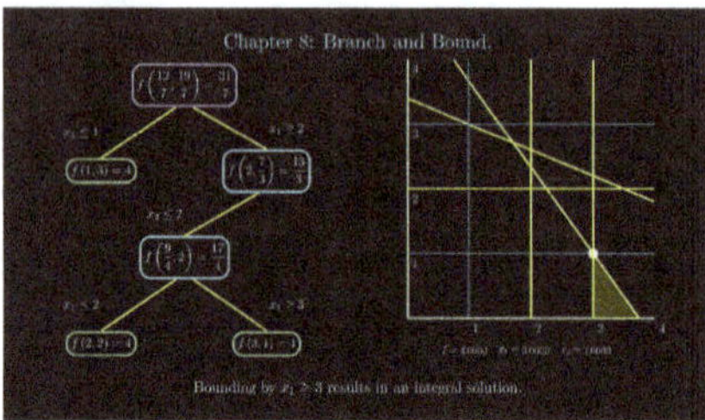

Figure 8.4.1 A demonstration of the Branch and Bound method.

The other more tableau centric approach is the **Gomory Cutting Plane** method Definition 8.2.4. We pivot to the optimal relaxed solution, and if this is nonintegral, we select a nonintegral b_i and for that row, add the additional constraint

$$\sum_j (\lfloor a_{ij} \rfloor - a_{ij})x_j \leq (\lfloor b_i \rfloor - b_i)$$

which excises the nonintegral optimal solution but preserves all integral solutions of the original problem. We repeat until an optimal solution is achieved.

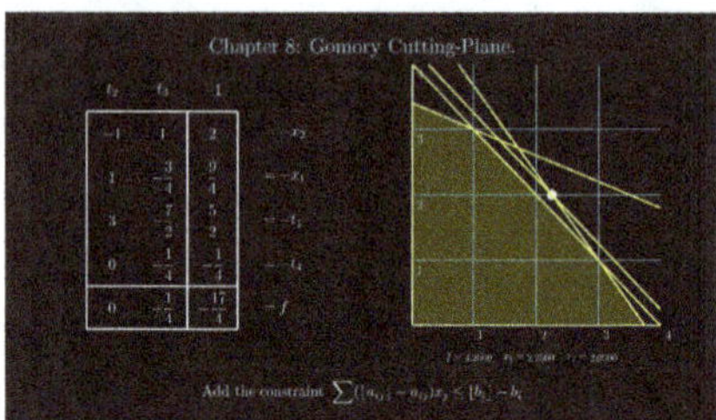

Figure 8.4.2 A demonstration of the Gomory Cutting Plane method.

8.5 Problems for Chapter 8

Exercises

1. For each of the linear optimization problems, solve the linear relaxation problem, then use the graphical method to find a solution if we restrict to integer values.

    (a)

$$\textbf{Maximize: } f(x_1, x_2) = 3x_1 + 2x_2$$
$$\textbf{subject to: } 2x_1 + 4x_2 \leq 15$$
$$5x_1 + 3x_2 \leq 22$$
$$x_1, x_2 \geq 0 \text{ and integral.}$$

    (b)

$$\textbf{Minimize: } g(x_1, x_2) = 10x_1 + x_2$$
$$\textbf{subject to: } 3x_1 - 2x_2 \geq 15$$
$$7x_1 + 4x_2 \leq 50$$
$$x_1, x_2 \geq 0 \text{ and integral.}$$

    (c)

$$\textbf{Maximize: } h(x_1, x_2) = 5x_1 + 4x_2$$
$$\textbf{subject to: } 3x_1 + 8x_2 \leq 13$$
$$4x_1 - 3x_2 \leq 25$$
$$x_1 \geq 0 \text{ and } x_1, x_2 \text{ are integral.}$$

2. Prove or find a counterexample: Let $\mathbf{x}$ be the solution to the linear relaxation for an integer optimization problem, such that $\mathbf{x}$ has only integer coordinates. Then $\mathbf{x}$ is a solution to the original integral problem.

3. Come up with 2 maximization problems, one two dimensional and one three dimensional, where only integer solutions are sensible, and explain why these problems should be integral problems. Then do the same for two minimization problems.

4. For each of the following integer optimization problems, find an integral solution using the branch and bound method, and using the cutting plane method.

    (a)

$$\textbf{Maximize: } f(x_1, x_2) = 5x_1 + 4x_2$$
$$\textbf{subject to: } 3x_1 + 4x_2 \leq 10$$
$$x_1, x_2 \geq 0 \text{ and integral.}$$

    (b)

$$\textbf{Minimize: } g(x_1, x_2) = 3x_1 + 6x_2$$
$$\textbf{subject to: } 7x_1 + 3x_2 \geq 40$$
$$x_1, x_2 \geq 0 \text{ and integral.}$$

(c)

$$\text{Maximize: } h(x_1, x_2) = x_1 + x_2$$
$$\text{subject to: } 3x_1 + 2x_2 \leq 5$$
$$x_2 \leq 2$$
$$x_1, x_2 \geq 0 \text{ and integral.}$$

(d)

$$\text{Maximize: } \alpha(x, y) = 2x + y$$
$$\text{subject to: } 3x + y \leq 13$$
$$-x + 2y \leq 6$$
$$x, y \geq 0 \text{ and integral.}$$

(e)

$$\text{Minimize: } \beta(x, y) = x - y$$
$$\text{subject to: } 3x + 4y \leq 6$$
$$x - y \leq 1$$
$$x, y \geq 0 \text{ and integral.}$$

(f)

$$\text{Maximize: } f(x, y) = 8x + 3y$$
$$\text{subject to: } 2x + y \leq 13$$
$$2x \leq 5$$
$$x, y \geq 0 \text{ and integral.}$$

(g)

$$\text{Maximize: } f(x_1, x_2, x_3) = -4x_1 + x_2 + 2x_3$$
$$\text{subject to: } 2x_1 + 3x_2 \leq 28$$
$$-x_1 + x_3 \leq 12$$
$$x_1, x_2, x_3 \geq 0 \text{ and integral.}$$

5. Solve the following integer optimization problems.

   (a) A potter makes sculptures and bowls out of clay. It takes 8 hours
   and 2 pounds of clay to make a sculpture, 2 hours and 2 pounds of
   clay for a bowl. She has 50 hours a week and 30 pounds of clay with
   which to make things. She can sell sculptures for $200 and bowl for
   $20. How much of each should she make to maximize revenue?

   (b) A man is preparing food for a party at his house, and is making sure
   there is enough. A chicken pot pie takes 300 g of flour and 450 g
   of chicken. He air and land wellington takes 240 g of flour, 150 g
   of chicken and 1000 g of beef. Evidently he knows no other recipes.
   He has 2400 g of flour, 2700 g of chicken and 6000 g of beef. The
   pot pies feed 2 people, the wellington feeds 8. How many of each
   should he make?

   (c) A family of 12 gnomes have three mines A, B and C from which
   to dig gems. A gnome digging in mine A can dig up 108 gems a

week, 75 gems a week in mine B and 120 gems a week in mine C. They have a budget of 75 gold pieces (gp) a month for operating expenses. A gnome digging in mine A has expenses of 7 gp a month, 5 gp a month in mine B, and 16 gp a month in mine C. Due to size limitations, at most 5 gnomes can dig in mine C. How should this family distribute the gnomes amongst the mines to maximize gem production?

6.

   (a) What do you think would happen if the Gomory Plane Cutting algorithm Definition 8.2.4 was applied to a linear optimization problem where the relaxed problem achieved optimality, but the integral restriction had no solutions?

   (b) Test your conjecture on the following problem:

$$\textbf{Maximize: } f(x, y) = 2x + y$$
$$\textbf{subject to: } 3x + 3y \leq 2$$
$$4y \geq 1$$
$$x, y \geq 0 \text{ and integral.}$$

7. Consider a primal integral linear maximization problem with objective function f such that the integral problem has an optimal solution $\mathbf{x}'$. Then for each of the following find a counterexample.

   (a) The linear relaxation of this primal problem also achieves optimality. (Think irrational numbers.)

   (b) Suppose that the linear relaxation also achieves optimality at $\mathbf{x}''$, then the integral dual to the original integral max problem must also achieve an optimal solution.

   (c) Suppose that the linear relaxation also achieves optimality at $\mathbf{x}''$ such that $f(\mathbf{x}') < f(\mathbf{x}'')$. Suppose the dual to the integral maximization problem achieved optimality at $\mathbf{y}'$, and let $\mathbf{y}''$ denote the dual to the relaxation. Then $g(\mathbf{y}') > g(\mathbf{y}'')$ where g is the objective function of the dual.

Chapter 9

Extra Topics

This is a chapter on assorted applications of linear optimization that are not generally a standard part of the undergraduate course, but are fun and nifty extensions. There is a good chance that anyone taking or teaching a class on linear optimization will not be able to cover these topics. However, for a class who finds the time, as independent reading for an ambitious or curious student, or fodder for projects or exploration, these can provide some insight into the vast array of problems to which the theory from this course may be applied.

In Section 9.1, we discuss graph matchings and coverings using linear optimization. In Section 9.2 linear optimization is used to solve sudoku and sudoku like puzzles. In Section 9.3 scheduling problems are modeled as and solved as linear optimization problems. Finally, Section 9.4 provides an alternative approach to the proof of the Strong Duality Theorem Theorem 4.2.4.

9.1 Coverings and Matchings of Graphs

Coverings and matchings on graphs are an integral part of the study of graph theory with numerous applications. A full exploration would be more appropriate for a graph theory or combinatorics course. However, to highlight some of the ways that linear optimization can be applied here, we examine the relationship between primal and dual problems, observe some limitations, and consider a case where we can employ linear optimization to solve problems without concern.

Exploration 9.1.1 Consider the following undirected graph G.

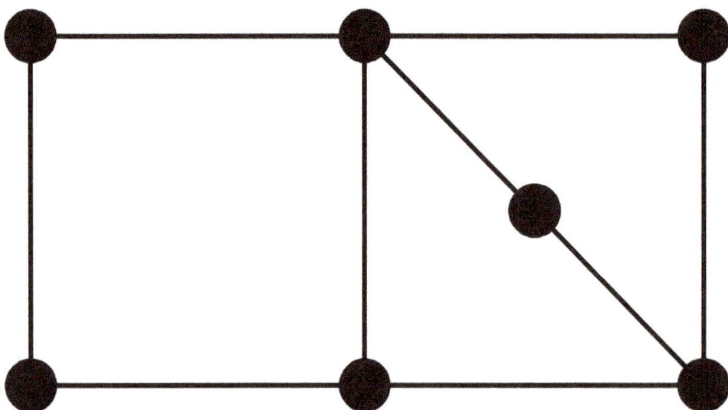

(a) A **matching** is a collection of edges M such that no two edges in M

are incident to the same vertex. Let $\Xi(G)$ denote the size of the largest possible matching(s) [1].

What is $\Xi(G)$? How can we be sure this is true?

(b) Find $\Xi(H)$ for H:

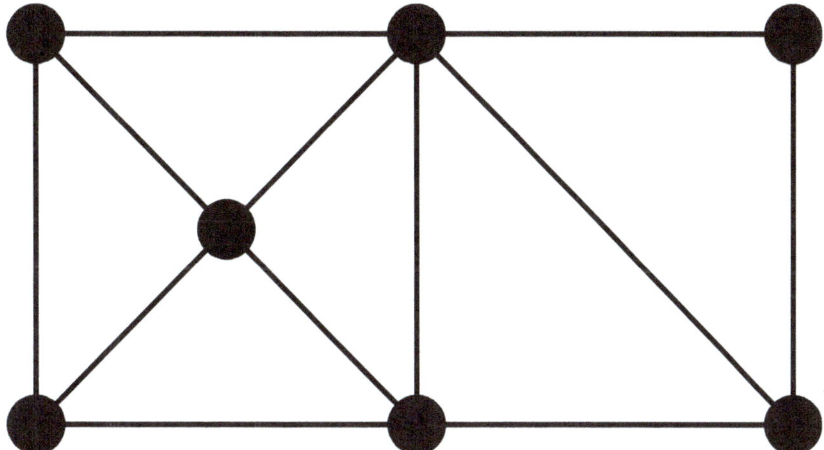

Activity 9.1.2 Let G be a graph with vertices $v_1, \ldots, v_n$, and M be any matching on G. For each edge $v_i v_j$ let $x_{ij} \in [0, 1]$ such that if $v_i v_j \in M$, then $x_{ij} = 1$, otherwise $x_{ij} = 0$.

We now construct a canonical maximization problem which can help compute a maximum matching.

(a) For each vertex v_i, write an inequality to ensure that M is a matching.

(b) Given the above constraints, do we need another constraint to ensure that $x_{ij} \leq 1$?

(c) Find a linear objective function to compute $\Xi(G)$.

(d) State the maximization linear optimization problem for computing the maximum matching of a graph G. We well refer to this problem as the **matching primal problem**.

(e) Consider G from Exploration 9.1.1. Label each vertex and write out a Tucker tableau for the linear optimization problem for computing the maximum matching.

(f) Solve the above optimization problem:

```
%display typeset
A = (FIX_ME)
b = (FIX_ME)
c = (FIX_ME)
P = InteractiveLPProblem(A, b, c,
  constraint_type = "<=",
  variable_type = ">=")
P
```

```
print(P.optimal_solution())
print(P.optimal_value())
```

---

[1]There is, too my knowledge, no standard notation for the size of a maximum matching for G. This is the proposed notation from Dr. Mark Kayll of the University of Montana, since the Ξ, the capital greek letter Xi, looks like a matching.

What do you notice?

(g) Consider H from Exploration 9.1.1. Label each vertex and write out a Tucker tableau for the linear optimization problem for computing the maximum matching.

(h) Solve the above optimization problem:

```
%display typeset
AH = (FIX_ME)
bH = (FIX_ME)
cH = (FIX_ME)
PH = InteractiveLPProblem(AH, bH, cH,
 constraint_type = "<=",
 variable_type = ">=")
PH
```

```
print(PH.optimal_solution())
print(PH.optimal_value())
```

What do you notice?

Activity 9.1.3 Consider the general maximization problem constructed in Activity 9.1.2.

(a) Prove that any graph G and matching M (maximum or not) corresponds to a feasible solution for this problem where $x_{ij} = 1$ if $v_i v_j \in M$ and $x_{ij} = 0$ otherwise.

(b) Let $f(\mathbf{x}^*)$ be the maximum value of the objective function for this problem. What can be said about the relationship between $f(\mathbf{x}^*)$ and $\Xi(G)$?

    A. $f(\mathbf{x}^*) \leq \Xi(G)$.   D. No general relationship exists
    B. $f(\mathbf{x}^*) \geq \Xi(G)$.   between $f(\mathbf{x}^*)$ and $\Xi(G)$.
    C. $f(\mathbf{x}^*) = \Xi(G)$.

Prove your claim.

Activity 9.1.4 Consider the general maximization problem constructed in Activity 9.1.2. We now consider its dual problem.

(a) Let y_i denote the dual variable corresponding to the primal constraint for vertex v_i. What is the dual objective function in terms of y_i?

(b) For each edge $v_i v_j$, there is a dual constraint, state this dual constraint. (Hint: in the original Tucker tableau, when would an entry in the x_{ij} column be a zero, and when would it be a one?)

(c) State the dual minimization problem to the primal maximum matching problem. We will refer to this problem as the **dual covering problem**.

(d) Suppose we restrict to only integer values, give an interpretation for the dual min problem (Hint: each y_i corresponds to a vertex v_i. Would y_i take on any values besides 0 or 1?)

Activity 9.1.5 Given a graph G, a **vertex cover** or *cover* is a collection of vertices U such that for any edge $v_i v_j$ either v_i or v_j (possibly both) are in U. Let $\tau(G)$ denote the size of the smallest vertex cover.

(a) Find $\tau(G)$ for G from, Exploration 9.1.1.

(b) Find $\tau(H)$ for H from, Exploration 9.1.1.

(c) For each vertex v_i, let $y_i = 1$ if $v_i \in U$ and $y_i = 0$ otherwise. Show this is a feasible solution to the dual problem found in Activity 9.1.4 for G. Is it optimal?

(d) Since we solved the matching problem for G in Exploration 9.1.1, use Sage to solve the dual problem

```
%display typeset
D = P.dual()
D
```

```
print(D.optimal_solution())
print(D.optimal_value())
```

What do we notice?

(e) We "solved" the matching maximization problem for H in Exploration 9.1.1, use Sage to solve the dual problem

```
%display typeset
DH = PH.dual()
DH
```

```
print(DH.optimal_solution())
print(DH.optimal_value())
```

What do we notice?

Activity 9.1.6 Consider the general dual minimization problem constructed in Activity 9.1.4.

(a) Prove that any graph G and cover U (minimum or not) corresponds to a feasible solution for this problem where $y_i = 1$ if $v_i \in U$ and $y_i = 0$ otherwise.

(b) Let $g(\mathbf{y}^*)$ be the minimum value of the objective function for this problem. What can be said about the relationship between $g(\mathbf{y}^*)$ and $\tau(G)$?

A. $g(\mathbf{y}^*) \leq \tau(G)$.

B. $g(\mathbf{y}^*) \geq \tau(G)$.

C. $g(\mathbf{y}^*) = \tau(G)$.

D. No general relationship exists between $g(\mathbf{y}^*)$ and $\tau(G)$.

Prove your claim.

(c) What can be said about the relationship between $\Xi(G)$ and $\tau(G)$?

A. $\Xi(G) \leq \tau(G)$.

B. $\Xi(G) \geq \tau(G)$.

C. $\Xi(G) = \tau(G)$.

D. No general relationship exists between $\Xi(G)$ and $\tau(G)$.

Prove your claim.

9.1.1 Königs Theorem and Bipartite Graphs

As mentioned above, a full discussion of covers and matchings, while fascinating, would be beyond the scope of this text. We will restrict ourselves to a specific family of graphs.

Definition 9.1.7 A graph G is said to be bipartite, if its vertices V may be partitioned into two disjoint sets, V_1, V_2 where there are no edges between vertices in the same V_i. $\Diamond$

Activity 9.1.8

(a) Is G from Exploration 9.1.1 bipartite?

(b) Is H from Exploration 9.1.1 bipartite?

(c) Which of the following are bipartite?

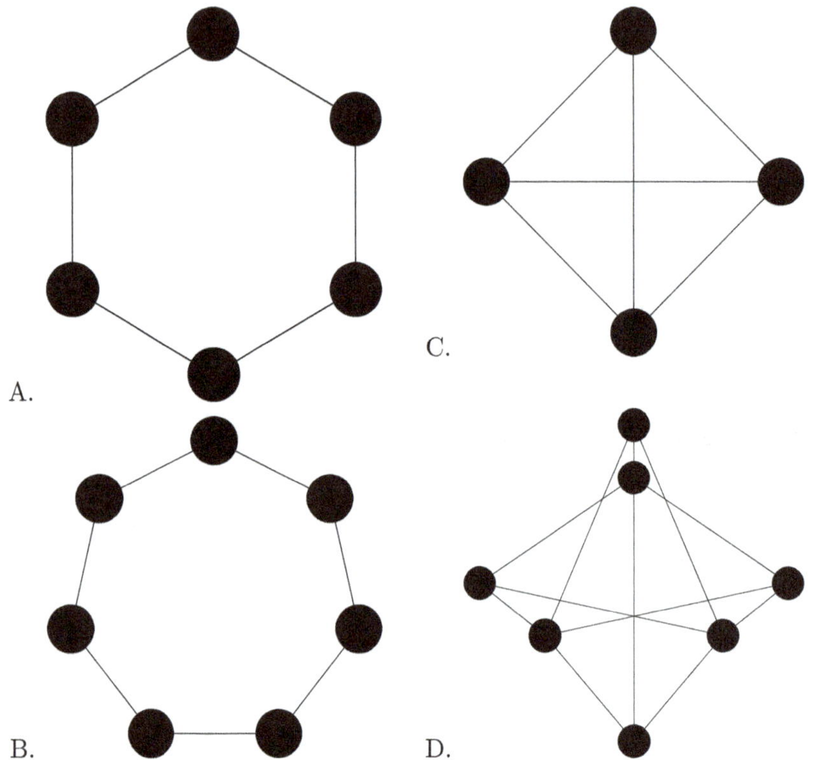

Activity 9.1.9 Prove that if a graph is bipartite, then it must *not* contain an odd length cycle [2].

We now consider coverings and matchings on only bipartite graphs.

Activity 9.1.10 For the graphs in Activity 9.1.8, find Ξ, τ. What do you notice? Is there any difference in the results for bipartite and nonbipartite graphs?

Activity 9.1.11 Consider the general linear optimization problems found in Activity 9.1.2 and Activity 9.1.4

(a) Show that if the primal matching problem has an optimal solution $\mathbf{x}^*$ consisting of only integer values, then it corresponds to a maximum matching.

---

[2]The converse is also true, but we will leave that alone here.

(b) Show that if the dual covering problem has an optimal solution $\mathbf{y}^*$ consisting of only integer values, then it corresponds to a minimum cover.

We now consider a general bipartite graph G, and we suppose the primal matching problem has an optimal solution $\mathbf{x}^*$ with potentially fractional values. We will explore how we can convert this solution into an integral valued optimal solution.

Activity 9.1.12 Let G be a bipartite graph, and let $\mathbf{x}^*$ be an optimal solution to the primal matching problem from Activity 9.1.2.

Suppose there were a collection of edges for which the corresponding x_{ij}^* had fractional values, such that these fractional edges formed a cycle C. Without loss of generality, we may label the vertices $v_1, v_2, \ldots v_k$ so that for $1 \le i \le k-1$, $v_i v_{i+1}$ and $v_k v_1$ have an edge between them, and $x_{i,i+1}^*, x_{k,1}^*$ have fractional values.

To make notation bearable, we'll understand that $v_{k+1} = v_1$.

(a) Why must k be even?

(b) Suppose we construct a new solution $\mathbf{x}'$ by replacing $x_{i,i+1}^*$ with $x_{i,i+1}^* - \epsilon$ when i is odd, with $x_{i,i+1}^* + \epsilon$ when i is even, and leaving every edge not part of C the same. What value for ϵ would guarantee that at least one of the new $x_{i,i+1}'$ is an integer?

    A. $\epsilon = \min\{x_{i,i+1}^* : 1 \le i \le k\}$.

    B. $\epsilon = \min\{x_{i,i+1}^* : 1 \le i \le k, i \text{ is odd}\}$.

    C. $\epsilon = \min\{x_{i,i+1}^* : 1 \le i \le k, i \text{ is even}\}$.

(c) Show that $\mathbf{x}' \ge 0$.

(d) Show that for any vertex v_j *not* a part of C, the bound corresponding to v_j is still satisfied by $\mathbf{x}'$.

(e) Show that for any vertex v_i part of C, the bound corresponding to v_i is still satisfied by $\mathbf{x}'$.

(f) Show that $f(\mathbf{x}') = f(\mathbf{x}^*)$.

(g) When comparing $\mathbf{x}^*$ and $\mathbf{x}'$, which solution has fewer integer values?

    A. $\mathbf{x}^*$ has fewer integer values.

    B. $\mathbf{x}'$ has fewer integer values.

    C. $\mathbf{x}^*, \mathbf{x}'$ have the same number of integer values.

    D. This cannot be determined.

Activity 9.1.13 Let G be a bipartite graph, and let $\mathbf{x}^*$ be an optimal solution to the primal matching problem from Activity 9.1.2.

Suppose there were *no* collection of edges for which the corresponding x_{ij}^* had fractional values, such that these fractional edges formed a cycle. Let $v_1, v_2, \ldots v_k$ form a maximal path P where $x_{i,i+1}^*, 1 \le i \le k - 1$ has fractional value. Note that v_1, v_k are the endpoints of P.

(a) Since P is maximal, any edges not a part of this path incident to v_1, v_k must be assigned an integer value. What must this value be?

(b) Suppose we construct a new solution $\mathbf{x}'$ by replacing $x_{i,i+1}^*$ with $x_{i,i+1}^* - \epsilon$ when i is odd, with $x_{i,i+1}^* + \epsilon$ when i is even, and leaving every edge not part of P the same. What value for ϵ would guarantee that at least one of the new $x_{i,i+1}'$ is an integer?

    A. $\epsilon = \min\{x^*_{i,i+1} : 1 \leq i \leq k\}$.

    B. $\epsilon = \min\{x^*_{i,i+1} : 1 \leq i \leq k, i \text{ is odd}\}$.

    C. $\epsilon = \min\{x^*_{i,i+1} : 1 \leq i \leq k, i \text{ is even}\}$.

(c) Show that $\mathbf{x}' \geq 0$.

(d) Show that for any vertex v_j *not* a part of P, the bound corresponding to v_j is still satisfied by $\mathbf{x}'$.

(e) Show that for any vertex v_i part of P, the bound corresponding to v_i is still satisfied by $\mathbf{x}'$.

(f) Show that $f(\mathbf{x}') \geq f(\mathbf{x}^*)$. Why does this imply $f(\mathbf{x}') = f(\mathbf{x}^*)$?

(g) When comparing $\mathbf{x}^*$ and $\mathbf{x}'$, which solution has fewer integer values?

    A. $\mathbf{x}^*$ has fewer integer values.

    B. $\mathbf{x}'$ has fewer integer values.

    C. $\mathbf{x}^*, \mathbf{x}'$ have the same number of integer values.

    D. This cannot be determined.

We now switch our attention to covers. Suppose the dual covering problem has an optimal solution $\mathbf{y}^*$ with potentially fractional values. We will explore how we can convert this solution into an integral valued optimal solution.

Activity 9.1.14 Let G be a bipartite graph, and let $\mathbf{y}^*$ be an optimal solution to the dual covering problem from Activity 9.1.4.

    Let $F \subseteq V$ be the set of vertices where y^*_i has a fractional value for all $v_i \in F$. Without loss of generality, suppose $V_1 \cap F \geq V_2 \cap F$.

(a) Let F^c denote the complement of F. We may partition $V(G)$ into four sets: $F \cap V_1, F \cap V_2, F^c \cap V_1, F^c \cap V_2$, Let edges of G be partitioned into four sets as follows:

- E_1 denotes edges incident to vertices in $F^c \cap V_1$ and $F^c \cap V_2$.
- E_2 denotes edges incident to vertices in $F \cap V_1$ and $F \cap V_2$.
- E_3 denotes edges incident to vertices in $F^c \cap V_1$ and $F \cap V_2$.
- E_4 denotes edges incident to vertices in $F \cap V_1$ and $F^c \cap V_2$.

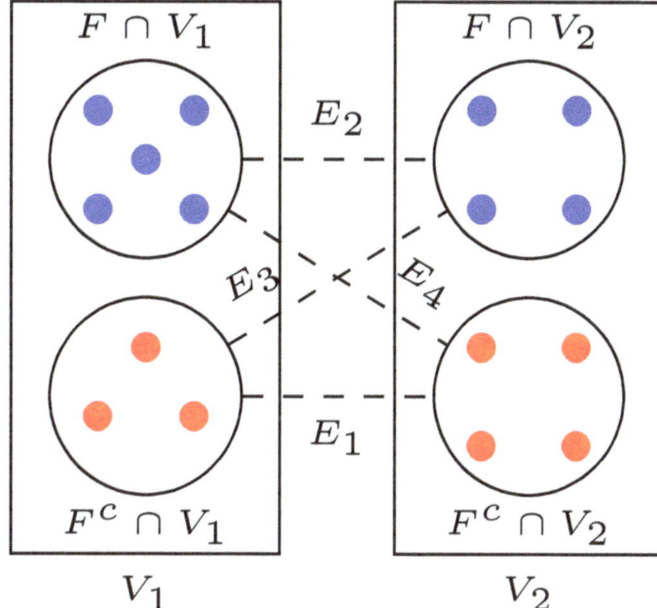

Recall that each vertex v_i in F is assigned a fractional value y_i^* less than 1. For any edge $v_i v_j$ in G, where $v_j \notin F$, what can we say about y_j^*?

(b) Recall that without loss of generality, $|F \cap V_1| \geq |F \cap V_2|$. Suppose we construct a new solution $\mathbf{y}'$ by replacing y_i^* with $y_i^* - \epsilon$ when $v_i \in F \cap V_1$, with $y_i^* + \epsilon$ when $v_i \in F \cap V_2$, and leaving every vertex not in F the same. What value for ϵ would guarantee that at least one of the new y_i' is an integer, and $\mathbf{y}' \geq 0$?

    A. $\epsilon = \min\{y_i^* : v_i \in V(G)\}$.

    B. $\epsilon = \min\{y_i^* : v_i \in V_1\}$.

    C. $\epsilon = \min\{y_i^* : v_i \in V_2\}$.

    D. $\epsilon = \min\{y_i^* : v_i \in F \cap V_1\}$.

    E. $\epsilon = \min\{y_i^* : v_i \in F \cap V_2\}$.

(c) For each edge in E_1, show that the corresponding constraint is still satisfied.

(d) For each edge in E_2, show that the corresponding constraint is still satisfied.

(e) For each edge in E_3, show that the corresponding constraint is still satisfied.

(f) For each edge in E_4, show that the corresponding constraint is still satisfied.

(g) Show that $g(\mathbf{y}') \leq g(\mathbf{y}^*)$. Why does this imply $g(\mathbf{y}') = g(\mathbf{y}^*)$?

(h) When comparing $\mathbf{y}^*$ and $\mathbf{y}'$, which solution has fewer integer values?

    A. $\mathbf{y}^*$ has fewer integer values.

    B. $\mathbf{y}'$ has fewer integer values.

    C. $\mathbf{y}^*, \mathbf{y}'$ have the same number of integer values.

    D. This cannot be determined.

We are finally ready to state our main result, König's Theorem.

Theorem 9.1.15 König's Theorem. *Let G be a bipartite graph. Then* $\Xi(G) = \tau(G)$.

Activity 9.1.16 Use Activity 9.1.11, Activity 9.1.12, Activity 9.1.13, and Activity 9.1.14 to prove Theorem 9.1.15.

9.2 Sudoku

In this section, we explore a curious application of linear optimization, solving sudokus!

Exploration 9.2.1 Consider the following sudoku puzzle:

6					4		1	
		7			1		2	
	3	8	9	2	7			
		6				7	9	4
5		9	7		2	6		
3	7	1		7		5		1
			5		9	8	4	
	6		2			1		
	8		6					9

Note that the rules for sudoku are that we fill in each entry with an integer $1, \ldots, 9$ so that:

- Each row contains exactly one of each number.

- Each column contains exactly one of each number.

- Each 3×3 "block" contains exactly one of each number.

(a) Solve the above sudoku if you feel like it.

Remark 9.2.2 We consider the standard sudokus like the one above an **order** sudoku: we have 3×3 "blocks", each with 3×3 entries. Potential values range from 1 through 3^2. Other orders are also possible.

More sudokus may be found at https://www.websudoku.com/

Activity 9.2.3 Consider a general order 3 sudoku puzzle. We want to define a linear maximization problem that solves a given puzzle. Let $x_{ijk} = 1$ is a solution has value k in entry i, j (measured from the bottom left), and $x_{ijk} = 0$ otherwise.

(a) For the i, jth entry, write a linear equality constraint which ensures a value is chosen.

(b) For row i, write 9 equality conditions so that this row contains one of each entry.

(c) For column j, write 9 equality conditions so that this column contains one of each entry.

(d) For an arbitrary block, write 9 equality conditions so that this block contains one of each entry.

(e) Now consider the sudoku puzzle from Exploration 9.2.1. How could we write out appropriate equality conditions fixing each entry?

(f) Are there any other constraints we need?

(g) What should the objective function be? (Does it matter?)

Obviously, this would be an absolutely absurdly large problem to even fully write out, much less solve. We consider something simpler.

Activity 9.2.4 Consider the following order 2 sudoku puzzle:

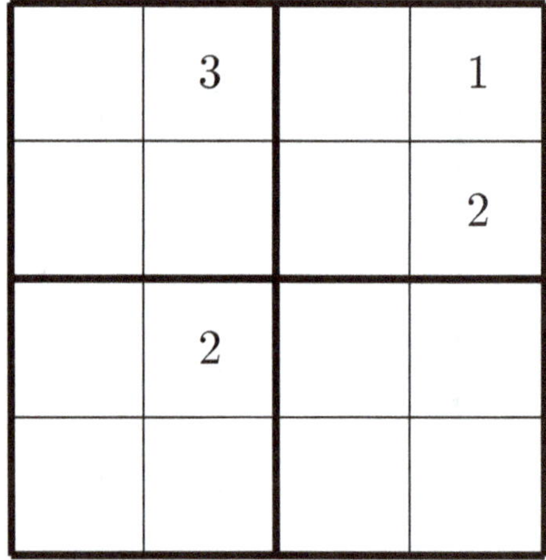

(a) Following Activity 9.2.3, write out the linear optimization problem which would compute the solved puzzle.

(b) If you *really* want to, solve this problem (there are 64 decision variables so...):

```
%display typeset
A = (FIXME)
b = (FIXME)
c = (FIXME)
P = InteractiveLPProblem(A, b, c,
  [FIXME],
  constraint_type = [FIXME],
  variable_type = "",
  problem_type = FIXME)
P
```

```
print(P.optimal_solution())
print(P.optimal_value())
```

This provides an opportunity for an enterprising student to engage in some further exploration.

Project 9.2.5 Write *efficient* code in Sage where one inputs a 9×9 matrix representing a sudoku puzzle (with maybe 0's for blank entries), and the code produces the appropriate linear optimization problem and solves it.

For more advanced or experienced coders, generalize this to allow the `order` of the sudoku puzzle to be a parameter.

9.3 Scheduling

In this section, we cover the problem of scheduling under time constraints and limited availability.

Exploration 9.3.1 Consider the last time you had to pick classes to fit a schedule. What were some things you had to consider in doing so? What sort of constraints did you consider necessary, and which were considered preferences? How does your answers compare to your classmates?

Activity 9.3.2 Yafa is an incoming freshman at Fantasi College and she is picking classes for her first semester. Fantasi College has classes either on Monday-Wednesday-Friday from 8:30-9:30, 10:00-11:00, 12:30-1:30 and 2:00-3:00, and on Tuesday-Thursday from 8:30-10:00, 10:30-12:00, and 1:00-2:30.

Yafa has put together a list of potential classes she could take this semester, and assigned to them a score from 1 - 10 depending on her interest in the class, the time of day, and her own "research" looking up professors on external rating sites. (Yafa has not yet taken her introductory statistics course, and so doesn't yet know how unreliable and biased these sites, and reviews in general, are.)

(Note that this is not altogether realistic, in that this is a massive oversimplification for the purpose of understanding the key ideas.)

# (j)	Course Name and Time	Score (c_j)	# (j)	Course Name and Time	Score (c_j)
1	Art MWF 12:30-1:30	8	14	Lit MWF 12:30-1:30	7
2	Art MWF 2:00-3:00	5	15	Lit MWF 2:00-3:00	6
3	Art TuTh 10:30-12:00	7	16	Lit TuTh 8:30-10:00	5
4	Art TuTh 1:00-2:30	5	17	MathA MWF 8:30-9:30	3
5	CS MWF 10:00-11:00	8	18	MathA MWF 10:00-11:00	5
6	CS TuTh 10:30-12:00	8	19	MathA TuTh 8:30-10:00	7
7	Econ MWF 8:30-9:30	6	20	MathB MWF 10:00-11:00	9
8	Econ MWF 12:30-1:30	7	21	MathB MWF 12:30-1:30	7
9	Econ TuTh 10:30-12:00	8	22	MathB TuTh 10:30-12:00	8
10	Econ TuTh 1:00-2:30	7	23	Sem MWF 10:00-11:00	6
11	Hist MWF 8:30-9:30	6	24	Sem MWF 10:00-11:00	8
12	Hist TuTh 10:30-12:00	8	25	Sem TuTh 10:30-12:00	8
13	Lit MWF 10:00-11:00	7	26	Sem TuTh 10:30-12:00	7

Additionally, she has the following stipulations to her schedule:

- No student is allowed to enroll in different offerings of the same class, MathA and MathB are considered different classes.

- No student can enroll in more than 1 course in the same time-slot.

- To prevent burnout, no student each student must have at least one time-slot off every day.

- No student can enroll in more than 6 or fewer than 4 classes.

- Each income student must enroll in a Freshman Seminar (Sem).

- Because of her major, Yafa intends to enroll in at least 1 Math class this semester (either A or B, not exclusive).

- Yafa wishes to get a head start on her general education, so she wants to take at least one of either Literature (Lit) or Art (Art). Similarly, she wishes to take at least one of Economics (Econ) or History (Hist).

Let x_j be 1 if she takes course j and 0 otherwise.

(a) What is a reasonable linear objective function to maximize?

(b) Write out the constraints which restrict students to at most 1 offering of the same class.

(c) Write out the constraints which restrict students to at most 1 course in a given time-slot.

(d) Write out the constraints which ensure each student will have at least one free time-slot a day.

(e) Write out the constraints which ensures each student is enrolled in at least 4 and at most 6 courses.

(f) Write out the constraint which ensures each student enrolls in one seminar.

(g) Write out the constraint which ensure that Yafa enrolls in (at least) one Math class.

(h) Write out the constraints which ensure that Yafa is on track with her general education credits.

(i) Write out the integer maximization problem we have constructed.

(j) (Optional) Solve it:

```
P = MixedIntegerLinearProgram(solver="GLPK")
X = P.new_variable(integer=True, nonnegative=True)
P.add_constraint(FIXME)
P.set_objective(FIXME)
P.show()
```

```
print(P.solve())
for i, v in sorted(P.get_values(X, convert=ZZ,
    tolerance=1e-3).items()):
    print(f'x_{i} = {v}')
```

Activity 9.3.3 Let's call Yafa from Activity 9.3.2 student 1. In a gross oversimplification suppose there were in total 5 students including Yafa registering, and that each class could sit at most 2 students.

Each student assigns their own score to each course, c_{ij} being the score that student i gives to course j. Let x_{ij} be 1 if student i enrolls in course j and 0 otherwise.

j	Course Name and Time	c_{1j}	c_{2j}	c_{3j}	c_{4j}	c_{5j}
1	Art MWF 12:30-1:30	8	10	4	6	9
2	Art MWF 2:00-3:00	5	8	5	5	7
3	Art TuTh 10:30-12:00	7	9	7	5	8
4	Art TuTh 1:00-2:30	5	6	3	5	6
5	CS MWF 10:00-11:00	8	2	8	9	5
6	CS TuTh 10:30-12:00	8	5	7	8	6
7	Econ MWF 8:30-9:30	6	6	8	4	7
8	Econ MWF 12:30-1:30	7	7	7	6	4
9	Econ TuTh 10:30-12:00	8	8	3	6	7
10	Econ TuTh 1:00-2:30	7	9	4	5	6
11	Hist MWF 8:30-9:30	6	5	7	7	6
12	Hist TuTh 10:30-12:00	8	6	8	9	7
13	Lit MWF 10:00-11:00	7	8	6	5	5
14	Lit MWF 12:30-1:30	7	8	7	6	7
15	Lit MWF 2:00-3:00	6	9	6	3	7
16	Lit TuTh 8:30-10:00	5	8	6	2	6
17	MathA MWF 8:30-9:30	3	2	5	5	6
18	MathA MWF 10:00-11:00	5	2	5	6	6
19	MathA TuTh 8:30-10:00	7	4	9	7	8
20	MathB MWF 10:00-11:00	9	1	1	1	8
21	MathB MWF 12:30-1:30	7	1	1	1	9
22	MathB TuTh 10:30-12:00	8	1	1	1	7
23	Sem MWF 10:00-11:00	6	10	9	7	4
24	Sem MWF 10:00-11:00	8	4	7	6	5
25	Sem TuTh 10:30-12:00	8	7	9	6	3
26	Sem TuTh 10:30-12:00	7	5	3	8	6

In addition to the constraints that all students have to abide by, we have the following constraint for each student:

- Yafa (student 1) still has her constraints from Activity 9.3.2.

- Student 2 has to take at least 1 Art class, either Computer Science (CS) or Math, and either Economics or History.

- Student 3 has to take at least 1 Economics class, 1 Math and 1 Literature class.

- Student 4 has to take a Computer Science class and a History class.

- Student 5 has to take both Math classes.

(a) How might the objective function and constraints from Activity 9.3.2 be adapted to this situation?

(b) What additional constraints are required?

(c) Depending on how you set up the objective function, a student who rates everything highly will be prioritized over a student who ranks everything more modestly. To prevent gaming the system, what could be done to ameliorate this effect?

(d) The Dean approves one of the classes to be increased in size to a whopping 3 students. How can the dual problem inform how this choice of class should be made?

Activity 9.3.4 Discuss how the discussions in Activity 9.3.2 and Activity 9.3.3 may be modified to apply to other situations.

Project 9.3.5 Now model a scheduling problem that is not oversimplified, perhaps one based on your current or former institution(s).

Project 9.3.6 For the truly ambitious, model a scheduling problem for the *offering of courses*, factoring in preferred times, preferred courses, preferred classrooms, changes in staffing, fluctuating student demand and flow of courses in the future etc.[1].

---

[1]This is as open a problem as there can be. Someone who solves this problem will gain more recognition than if they solved all the Millennium Problems.

9.4 Another Approach to Strong Duality

In this section, we provide an alternative approach to proving the Strong Duality Theorem Theorem 4.2.4.

Theorem 9.4.1 Hyperplane Seperation Theorem. *Given two disjoint convex sets $U, V \subseteq \mathbb{R}^n$, there is a hyperplane $H = \{\mathbf{x} \in \mathbb{R}^n : \mathbf{h}^\top \mathbf{x} = k\}$ for some $\mathbf{h} \in \mathbb{R}^n, k \in \mathbb{R}$, such that $\mathbf{h}^\top U \geq k, \mathbf{h}^\top V < k$.*

Activity 9.4.2

(a) Sketch two nonempty convex sets U, V, what does H look like here?

(b) Sketch two nonconvex sets where Theorem 9.4.1 fails.

Activity 9.4.3 We prove the case of Theorem 9.4.1 where there are $u_0 \in U, v_0 \in V$ that minimize $\{\|u - v\| : u \in U, v \in V\}$. We assume this is true.

(a) Without loss of generality, let $v_0 = 0$. Why can we do this?

(b) Let $\mathbf{h} = u_0 - v_0$. Sketch $u_0, v_0, \mathbf{h}$ and $H := \{\mathbf{x} \in \mathbb{R}^n : \mathbf{h}^\top \mathbf{x} = 0\}$.

(c) We want to show that H is the separating hyperplane. Suppose that U was not contained in $\{\mathbf{x} \in \mathbb{R}^n : \mathbf{h}^\top \mathbf{x} \geq 0\}$ what must be true about U?

(d) Let $u' \in U$ such that $\mathbf{h}^\top u' \leq 0$. Sketch u'.

(e) Let $\mathbf{d} = u' - u_0$ describe geometrically what $\frac{-u_0^\top \mathbf{d}}{\mathbf{d}^\top \mathbf{d}} \cdot \mathbf{d} + u$ represents. (Think dot product, projections and for $\frac{-u_0^\top \mathbf{d}}{\mathbf{d}^\top \mathbf{d}} \cdot \mathbf{d} + u$ to start at u.)

(f) Let $\alpha = \frac{-u_0^\top \mathbf{d}}{\mathbf{d}^\top \mathbf{d}}$, show that $0 < \alpha < 1$.

(g) Let $w = \frac{-u_0^\top \mathbf{d}}{\mathbf{d}^\top \mathbf{d}} \cdot \mathbf{d} + u_0$, show that $w = \alpha u' + (1 - \alpha)u_0$

(h) Show that $w \in U$.

(i) Show that $w^\top w = u_0^\top u_0 + \alpha \mathbf{d}^\top \mathbf{d} \left(2 \frac{u_0^\top \mathbf{d}}{\mathbf{d}^\top \mathbf{d}} + \alpha \right)$, and explain why $w^\top w < u_0^\top u_0$.

(j) Why does the last statement result in a contradiction?

We now introduce a key idea which will tie together the primal and dual problems.

Definition 9.4.4 Let $S \subseteq \mathbb{R}^n$. We call the **cone** of S, denoted $\mathrm{cone}(S)$ to be the set $\mathrm{cone}(S) := \left\{ \sum_{i=1}^{m} u_i \mathbf{s}_i : u_i \in \mathbb{R}, u_i \geq 0, \mathbf{s}_i \in S \right\}$. $\diamond$

Activity 9.4.5

(a) In $\mathbb{R}^2$, describe $\mathrm{cone}\left(\left\{ \begin{bmatrix} 1 \\ 1 \end{bmatrix}, \begin{bmatrix} 1 \\ -2 \end{bmatrix} \right\} \right)$.

(b) Prove that for any $S \subseteq \mathbb{R}^n, \mathrm{cone}(S)$ is convex.

(c) Let A denote a $n \times m$ matrix, and let P denote the cone of the columns of A.

Suppose $\mathbf{b} \notin P$. What does Theorem 9.4.1 tell us?

We can codify the intuition gained from Activity 9.4.5 in the following:

Theorem 9.4.6   The Farkas Lemma. *Given a matrix $A \in \mathbb{R}^{n \times m}$ and vector $\mathbf{b} \in \mathbb{R}^n$, exactly one of the following is true:*

A. *There is a $\mathbf{x} \in \mathbb{R}^m$ such that $\mathbf{x} \geq \mathbf{0}$ and $A\mathbf{x} = \mathbf{b}$.*

B. *There is a $\mathbf{y} \in \mathbb{R}^n$ such that $\mathbf{y}^\top \mathbf{b} < 0$ and $\mathbf{y}^\top A \geq \mathbf{0}$.*

Activity 9.4.7 We prove Theorem 9.4.6 and a useful corollary.

(a) Recall Activity 9.4.5 (c). How do the cases (A) and (B) of Theorem 9.4.6 relate to the cone P?

(b) Suppose both cases (A) and (B) of Theorem 9.4.6 held at the same time. use the product $\mathbf{y}^\top A\mathbf{x}$ to derive a contradiction.

(c) If Theorem 9.4.6 (A) were true, what could we say about (B)?

(d) Suppose Theorem 9.4.6 (A) were false. We want to prove that (B) is true. What does Activity 9.4.5 (c) tell us?

(e) Let's denote the normal vector of the separating hyperplane by $\mathbf{y}$ (interesting choice) so that $\mathbf{y}^\top \mathbf{b} < \mathbf{y}^\top p$ for any $p \in P$. Why must $\mathbf{y}^\top \mathbf{b} < 0$?

(f) Suppose A had a column A_j such that $\mathbf{y}^\top A_j < 0$, show that there is an $\alpha > 0$ such that $\mathbf{y}^\top (\alpha A_j) < \mathbf{y}^\top \mathbf{b}$. Why is this a contradiction?

(g) Why is Theorem 9.4.6 (B) proven? Why is then Theorem 9.4.6 proven?

(h) Now that Theorem 9.4.6 is proven, we apply it to $\begin{bmatrix} A & I_n \end{bmatrix}$ and $\mathbf{b}$.

Suppose (A) held, and we had that there was a $\begin{bmatrix} \mathbf{x} \\ \mathbf{t} \end{bmatrix} \geq 0$ so that

$$\begin{bmatrix} A & I_n \end{bmatrix} \begin{bmatrix} \mathbf{x} \\ \mathbf{t} \end{bmatrix} = \mathbf{b}.$$ How would $A\mathbf{x}$ compare to $\mathbf{b}$?

A. $A\mathbf{x} \leq \mathbf{b}$.
B. $A\mathbf{x} \geq \mathbf{b}$.
C. $A\mathbf{x} = \mathbf{b}$.

D. No comparison may be made between $A\mathbf{x}$ and $\mathbf{b}$.

(i) Suppose (A) failed. Then there is a $\mathbf{y} \in \mathbb{R}^n$ satisfing (B) for $\begin{bmatrix} A & I_n \end{bmatrix}$, $\mathbf{b}$.

What can we say about $\mathbf{y}$ compared to $\mathbf{0}$?

A. $\mathbf{y}^\top \leq \mathbf{0}^\top$.
B. $\mathbf{y}^\top \geq \mathbf{0}^\top$.
C. $\mathbf{y}^\top = \mathbf{0}^\top$.

D. No comparison may be made between $\mathbf{y}$ and $\mathbf{0}p$.

What can we say about $\mathbf{y}^\top A$ compared to $\mathbf{0}$?

A. $\mathbf{y}^\top A \leq \mathbf{0}^\top$.
B. $\mathbf{y}^\top A \geq \mathbf{0}^\top$.
C. $\mathbf{y}^\top A = \mathbf{0}^\top$.

D. No comparison may be made between $\mathbf{y}^\top A$ and $\mathbf{0}^\top$.

What can we say about $\mathbf{y}^\top \mathbf{b}$ compared to 0?

> A. $\mathbf{y}^\top \mathbf{b} \le 0$.
> B. $\mathbf{y}^\top \mathbf{b} \ge 0$.
> C. $\mathbf{y}^\top \mathbf{b} = 0$.
>
> D. No comparison may be made between $\mathbf{y}^\top \mathbf{b}$ and 0.

Corollary 9.4.8 The Farkas Lemma v2. *Given a matrix $A \in \mathbb{R}^{n \times m}$ and vector $\mathbf{b} \in \mathbb{R}^n$, exactly one of the following is true:*

A. *There is a $\mathbf{x} \in \mathbb{R}^m$ such that $\mathbf{x} \ge \mathbf{0}$ and $A\mathbf{x} \le \mathbf{b}$.*

B. *There is a $\mathbf{y} \in \mathbb{R}^n$ such that $\mathbf{y} \ge \mathbf{0}, \mathbf{y}^\top \mathbf{b} < 0$ and $\mathbf{y}^\top A \ge \mathbf{0}$.*

Without loss of generality, we may let $\mathbf{y}^\top \mathbf{b} = -1$ in case (B).

Recall the Strong Duality Theorem Theorem 4.2.4.

Theorem 9.4.9 The Strong Duality Theorem. *Given a pair of primal max-dual min problems, the primal max problem as an optimal solution $\mathbf{x}^*$ if and only if the dual min problem has an optimal solution $\mathbf{y}^*$. Moreover, $f(\mathbf{x}^*) = g(\mathbf{y}^*)$.*

Activity 9.4.10 Proof of the Strong Duality Theorem. We provide an alternate proof of Theorem 4.2.4 in this activity.

(a) Suppose that optimal dual solution $\mathbf{y}^*$ exists. Explain why by Activity 4.2.2 it suffices to show that $f(\mathbf{x}^*) \ge g(\mathbf{y}^*)$ for some feasible $\mathbf{x}^*$.

(b) Without loss of generality, let $d = 0$ and let $g^* = (\mathbf{y}^*)^\top \mathbf{b}$.

Consider the matrix $M = \begin{bmatrix} -\mathbf{c}^\top \\ A \end{bmatrix}$, and vector $\mathbf{d} = \begin{bmatrix} -g^* \\ \mathbf{b} \end{bmatrix}$. Apply Corollary 9.4.8 to $M, \mathbf{d}$. What does it mean for (A) to hold?

(c) If Corollary 9.4.8 (A) holds for $M, \mathbf{d}$, then it holds for the pair $-\mathbf{c}^\top, \begin{bmatrix} -g^* \end{bmatrix}$, as well as the pair $A, \mathbf{b}$ for the same $\mathbf{x} \in \mathbb{R}^m$. Why does this show that a feasible optimal solution $\mathbf{x}^*$ exists and that $f(\mathbf{x}^*) = g(\mathbf{y}^*)$?

(d) On the other hand, suppose Corollary 9.4.8 (B) holds for $M, \mathbf{d}$. What would it mean for (B) to hold?

(e) We would like to derive a contradiction.

Let $\begin{bmatrix} y_0 & \mathbf{y}^\top \end{bmatrix}$ denote the vector produced by Corollary 9.4.8 (B). Suppose $y_0 = 0$. How would $(\mathbf{y}^* + \mathbf{y})^\top A$ compare to $(\mathbf{y}^*)^\top A$ and $(\mathbf{y}^* + \mathbf{y})^\top \mathbf{b}$ compare to $(\mathbf{y}^*)^\top \mathbf{b}$? Why is this a contradiction?

(f) Suppose $y_0 > 0$. Let $\mathbf{y}' := \frac{\mathbf{y}}{y_0}$.

Show that since $\begin{bmatrix} y_0 & \mathbf{y}^\top \end{bmatrix} M \ge 0$ that $(\mathbf{y}')^\top A \ge \mathbf{c}^\top$.

(g) Show that since $\begin{bmatrix} y_0 & \mathbf{y}^\top \end{bmatrix} \mathbf{d} = -1$ that $(\mathbf{y}')^\top \mathbf{b} = g^* - \frac{1}{y_0}$.

(h) Explain why **(f)** and **(g)** produce a contradiction.

Appendix A

Review Material

A.1 Linear Algebra Review

This is an extremely brief review of linear algebra. It is understood that linear algebra is a pre-requisite for this course. However, everyone needs refreshers or a reference for specifics from time to time.

If a more thorough treatment is needed, then there are numerous linear algebra texts, and many that are OERs like this text. "Understanding Linear Algebra" by David Austin is an excellent text with a focus on developing geometric intuition, and less so on formal proofs. For a more theory oriented text, "Linear Algebra" by Jim Hefferon is an excellent choice.

Definition A.1.1 A real-valued **matrix** is a rectangular array of the form

$$
A = \begin{bmatrix}
a_{11} & a_{12} & \cdots & a_{1m} \\
a_{21} & a_{22} & \cdots & a_{2m} \\
\vdots & \vdots & \ddots & \vdots \\
a_{n1} & a_{n2} & \cdots & a_{nm}
\end{bmatrix}.
$$

Also denoted $A = [a_{ij}]_{n \times m}$ is a $n \times m$ matrix, denoting that A has n rows and m columns. We note that a_{ij} is the entry of A in row i, column j. $\Diamond$

Definition A.1.2 $n \times 1$ matrices are also referred to as **vectors**:

$$
\mathbf{x} = \begin{bmatrix}
x_1 \\
x_2 \\
\vdots \\
x_n
\end{bmatrix}.
$$

This is the convention we use, with vectors being column matrices. Some texts default to row vectors. $\Diamond$

Definition A.1.3 Given a $n \times m$ matrix A, we define the **transpose** of A

denoted $A^\top$ as $A = [a_{ij}]_{n \times m}^\top = [a_{ji}]_{m \times n}$ or

$$
\begin{bmatrix}
a_{11} & a_{12} & \cdots & a_{1m} \\
a_{21} & a_{22} & \cdots & a_{2m} \\
\vdots & \vdots & \ddots & \vdots \\
a_{n1} & a_{n2} & \cdots & a_{nm}
\end{bmatrix}^\top
=
\begin{bmatrix}
a_{11} & a_{21} & \cdots & a_{n1} \\
a_{12} & a_{22} & \cdots & a_{n2} \\
\vdots & \vdots & \ddots & \vdots \\
a_{1m} & a_{2m} & \cdots & a_{nm}
\end{bmatrix}.
$$

$\Diamond$

Example A.1.4

$$
\begin{bmatrix}
1 & -2 & 3 \\
0 & 5 & 6
\end{bmatrix}^\top
=
\begin{bmatrix}
1 & 0 \\
-2 & 5 \\
3 & 6
\end{bmatrix}.
$$

$\square$

Definition A.1.5 Given two matrices of the same dimensions $A = [a_{ij}]_{n \times m}, B = [b_{ij}]_{n \times m}$, we define their **sum** entrywise, that is: $A + B = [a_{ij} + b_{ij}]_{n \times m}$. $\Diamond$

Example A.1.6

$$
\begin{bmatrix}
1 & -2 & 3 \\
0 & 5 & 6
\end{bmatrix}
+
\begin{bmatrix}
4 & 7 & 0 \\
-8 & 2 & -4
\end{bmatrix}
$$

$$
=
\begin{bmatrix}
1+4 & -2+7 & 3+0 \\
0+(-8) & 5+2 & 6+(-4)
\end{bmatrix}
$$

$$
=
\begin{bmatrix}
5 & 5 & 3 \\
-8 & 7 & 2
\end{bmatrix}.
$$

$\square$

Definition A.1.7 Given matrices $A = [a_{ij}]_{n \times m}, B = [b_{ij}]_{m \times \ell}$, we define their **product** to be $AB = [c_{ij}]_{n \times \ell} = [\sum_{k=1}^{m} a_{ik} b_{kj}]_{n \times \ell}$. $\Diamond$

Example A.1.8

$$
\begin{bmatrix}
4 & 7 & 0 \\
-8 & 2 & -4
\end{bmatrix}
\begin{bmatrix}
1 & 0 \\
-2 & 5 \\
3 & 6
\end{bmatrix}
$$

$$
=
\begin{bmatrix}
4(1) + 7(-2) + 0(3) & 4(0) + 7(5) + 0(6) \\
-8(1) + 2(-2) + (-4)(3) & -8(0) + 2(5) + (-4)(6)
\end{bmatrix}
$$

$$
=
\begin{bmatrix}
5 & 5 & 3 \\
-8 & 7 & 2
\end{bmatrix}.
$$

$$
\begin{bmatrix}
1 & 0 \\
-2 & 5 \\
3 & 6
\end{bmatrix}
\begin{bmatrix}
4 & 7 & 0 \\
-8 & 2 & -4
\end{bmatrix}
$$

$$
= \begin{bmatrix}
1(4) + 0(-8) & 1(7) + 0(2) & 1(0) + 0(-4) \\
(-2)(4) + 5(-8) & (-2)(7) + 5(2) & (-2)(0) + 5(-4) \\
3(4) + 6(-8) & 3(7) + 6(2) & 3(0) + 6(-4)
\end{bmatrix}
$$

$$
= \begin{bmatrix}
4 & 7 & 0 \\
-48 & -4 & -20 \\
-36 & 33 & -24
\end{bmatrix}.
$$

$\square$

Note that this dry and technical presentation fails to capture even an iota of the beautiful and deep theory this operation is meant to encapsulate. Nor is it meant to. Please see the aforementioned texts for a deeper and richer discussion.

Definition A.1.9 Given a matrix $A = [a_{ij}]_{n \times m}$ and real number c, we define the **scalar product** to be $cA = [ca_{ij}]_{n \times m}$. $\diamond$

Example A.1.10

$$
3 \begin{bmatrix}
1 & 0 \\
-2 & 5 \\
3 & 6
\end{bmatrix}
= \begin{bmatrix}
3(1) & 3(0) \\
3(-2) & 3(5) \\
3(3) & 3(6)
\end{bmatrix}
$$

$$
= \begin{bmatrix}
3 & 0 \\
-6 & 15 \\
9 & 18
\end{bmatrix}.
$$

$\square$

Definition A.1.11 We denote the **zero matrix** as $\mathbf{0}_{n \times m} := [0]_{n \times m}$ or $\mathbf{0}$ if the dimensions are clear from context. $\diamond$

Theorem A.1.12 *For matrices A, B, C, D, and scalars c, d, assuming appropriate dimensions, the following hold.*

- $A + B = B + A$.
- $(A + B) + C = A + (B + C)$.
- $c(A + B) + C = cA + cB$.
- $(c + d)A = cA + dA$.
- $(cd)A = c(dA)$.
- $1A = A$.
- $0A = \mathbf{0}$.
- $A + \mathbf{0} = \mathbf{0} + A = A$.
- $A + (-A) = (-A) + A = 0$.

- $(AB)C = A(BC)$.
- $A(B + C) = AB + AC$.
- $(A + B)C = AC + BC$.
- $c(AB) = A(cB) = (cA)B$.
- $\mathbf{0}A = A\mathbf{0}$.
- $(A^\top)^\top = A$.
- $(A + B)^\top = A^\top + B^\top$.
- $(cA)^\top = c(A^\top)$.
- $(AB)^\top = B^\top A^\top$.

Definition A.1.13 $A = [a_{ij}]_{n \times n}$ is a **square matrix**. The entries where $i = j$ are the **diagonal** of A. If $a_{ij} = 0$ when $i \neq j$, then A is a **diagonal matrix**. $\diamond$

Definition A.1.14 The **identity matrix** I_n is the $n \times n$ diagonal square matrix where the diagonal entries are all 1. $\diamond$

Theorem A.1.15 *For A a $n \times n$ matrix, $AI_n = I_n A = A$.*

Definition A.1.16 For A a $n \times n$ matrix, we say A is invertible if there exists a $n \times n$ matrix B such that $AB = BA = I_n$. We usual call B the **inverse** of A and denote it A^{-1}. $\diamond$

Theorem A.1.17 *If A is an invertible square matrix, then A^{-1} is unique.*

Example A.1.18 If $A = \begin{bmatrix} 3 & -1 \\ -2 & 4 \end{bmatrix}$ then $A^{-1} = \begin{bmatrix} \frac{2}{5} & \frac{1}{10} \\ \frac{1}{5} & \frac{3}{10} \end{bmatrix}$, one can check this. $\square$

Note that not every matrix is invertible. For example $\begin{bmatrix} 3 & -1 \\ -6 & 2 \end{bmatrix}$ is not invertible.

Definition A.1.19 Let a set V be equipped with operations $+$ and a scalar product. Let $\mathbf{x}, \mathbf{y}, \mathbf{z} \in V$ and a, b be scalars. Then V is a **vector space** if it satisfies the following axioms:

1. **Associativity of vector addition:** $(\mathbf{x} + \mathbf{y}) + \mathbf{z} = \mathbf{x} + (\mathbf{y} + \mathbf{z})$.

2. **Commutativity of vector addition:** $\mathbf{x} + \mathbf{y} = \mathbf{y} + \mathbf{x}$.

3. **Identity element of vector addition:** there exists a vector $\mathbf{0}$ called the **zero vector** such that $\mathbf{0} + \mathbf{x} = \mathbf{x} + \mathbf{0} = \mathbf{x}$.

4. **Inverse elements of vector addition:** for each vector $\mathbf{x}$, there exists a vector $-\mathbf{x}$ called the **additive inverse** of $\mathbf{x}$ such that $-\mathbf{x} + \mathbf{x} = \mathbf{x} + (-\mathbf{x}) = \mathbf{0}$.

5. Compatibility of scalar multiplication with real multiplication: $(ab)\mathbf{x} = a(b\mathbf{x})$.

6. **Identity element of scalar multiplication:** $1\mathbf{x} = \mathbf{x}$.

7. **Distributivity of scalar multiplication with respect to vector addition:** $a(\mathbf{x} + \mathbf{y}) = a\mathbf{x} + a\mathbf{y}$.

8. **Distributivity of scalar multiplication with respect to field addition:** $(a + b)\mathbf{x} = a\mathbf{x} + b\mathbf{x}$.

$\diamond$

There are a wide variety of interesting vector spaces spanning across all subfields of math. However, for our purposes, we will stick to boring ol' $\mathbb{R}^n$.

Definition A.1.20 Let V be a vector space, then $W \subseteq V$ is a subspace of V if it is also a vector space, with the same operations. $\diamond$

Theorem A.1.21 *Let V be a vector space, then $W \subseteq V$ is a subspace of V, if W is nonempty, and if for any $\mathbf{p}, \mathbf{q} \in W$ and scalars a, b, we have that $a\mathbf{p} + b\mathbf{q} \in W$.*

Example A.1.22 In $\mathbb{R}^4$, the set

$$W = \left\{ \begin{bmatrix} u \\ 0 \\ w \\ 0 \end{bmatrix} : u, w \in \mathbb{R} \right\}$$

forms a subspace of $\mathbb{R}^4$. $\qquad\square$

Definition A.1.23 Let V be a vector space and $S \subseteq V$. Then a **linear combination** of these vectors is a sum:

$$\sum a_i \mathbf{s}_i, \text{ where } a_i \in \mathbb{R}, \mathbf{s}_i \in S$$

$\diamond$

Definition A.1.24 Let V be a vector space and $S \subseteq V$. Then the **span** of S defined

$$\text{span}(S) = \left\{ \sum a_i \mathbf{s}_i : a_i \in \mathbb{R}, \mathbf{s}_i \in S \right\}.$$

If $\text{span}(S) = V$ we say that S **spans** V. $\qquad\diamond$

Example A.1.25 The set

$$S = \left\{ \begin{bmatrix} 1 \\ 2 \\ 3 \end{bmatrix}, \begin{bmatrix} 2 \\ 5 \\ -4 \end{bmatrix}, \begin{bmatrix} 3 \\ 1 \\ 2 \end{bmatrix}, \begin{bmatrix} 2 \\ -3 \\ 19 \end{bmatrix} \right\}$$

spans $\mathbb{R}^3$. $\qquad\square$

Definition A.1.26 Let V be a vector space and $S \subseteq V$. Then S is **linearly independent** if the equation

$$\sum_{\mathbf{s} \in S} a_i \mathbf{s}_i = \mathbf{0}$$

if and only if each $a_i = 0$. Otherwise, S is dependent. $\qquad\diamond$

Example A.1.27 Since

$$3 \begin{bmatrix} 1 \\ 2 \\ 3 \end{bmatrix} + (-2) \begin{bmatrix} 2 \\ 5 \\ -4 \end{bmatrix} + (1) \begin{bmatrix} 3 \\ 1 \\ 2 \end{bmatrix} + (1) \begin{bmatrix} 2 \\ -3 \\ 19 \end{bmatrix} = \begin{bmatrix} 0 \\ 0 \\ 0 \end{bmatrix}$$

so $\left\{ \begin{bmatrix} 1 \\ 2 \\ 3 \end{bmatrix}, \begin{bmatrix} 2 \\ 5 \\ -4 \end{bmatrix}, \begin{bmatrix} 3 \\ 1 \\ 2 \end{bmatrix}, \begin{bmatrix} 2 \\ -3 \\ 19 \end{bmatrix} \right\}$ is dependent.

$\left\{ \begin{bmatrix} 1 \\ 2 \\ 3 \end{bmatrix}, \begin{bmatrix} 2 \\ 5 \\ -4 \end{bmatrix}, \begin{bmatrix} 3 \\ 1 \\ 2 \end{bmatrix} \right\}$ is linearly independent. $\qquad\square$

Theorem A.1.28 *Let V be a vector space.*

1. *Any superset of a spanning set of V is also a spanning set.*

2. *Any subset of a linearly independent set of vectors in V is also linearly independent.*

Definition A.1.29 Let V be a vector space and $B \subseteq V$. Then B is a **basis** of V if for any $\mathbf{x} \in V$,

$$\sum_{\mathbf{b} \in B} a_i \mathbf{b}_i = \mathbf{x}$$

always has a unique solution. $\Diamond$

Theorem A.1.30 *Let V be a vector space. A spanning, linearly independent subset of V is a basis of V.*

Example A.1.31

$$\left\{ \begin{bmatrix} 1 \\ 2 \\ 3 \end{bmatrix}, \begin{bmatrix} 2 \\ 5 \\ -4 \end{bmatrix}, \begin{bmatrix} 3 \\ 1 \\ 2 \end{bmatrix} \right\}$$

is a basis for $\mathbb{R}^3$. $\square$

A.2 Probability Theory Review

This is an extremely brief "review" of the limited probability theory we utilize in Chapter 5. It's not even particularly fair to call this a review, since probability is not a prerequisite to this course. However, the limited amount we use is fairly straightforward and intuitive.

If a more thorough treatment is needed, then depending on your goals, there are good options available. For someone looking to explore some elementary probability theory, the introductory statistics textbook "OpenIntro Statistics" by David Diez, Christopher Barr, and Mine Çetinkaya-Rundel does a good job presenting this material. It also is an excellent introductory statistics text with labs and data available. For a calculus-based, theory heavy treatment of this subject, I recommend "Probability: Lecture and Labs" by Mark Huber.

Definition A.2.1 In probability, an **experiment** is an occurrence with a measurable result. Each instance of an experiment is a **trial**. The possible results of each trial are called **outcomes**. The set of all possible outcomes for an experiment is the **sample space** for that experiment. ◇

Definition A.2.2 Given an experiment with sample space S:

- An **event** A is a subset of S.

- Supposing each outcome in the sample space is equally likely, then the **probability** of A, denoted $P(A)$ is

$$P(A) = \frac{|A|}{|S|}.$$

◇

Remark A.2.3 But what does it mean for an event A to have probability $P(A)$? It means that if I repeat the experiment over and over, the proportion of them where A is true should be $P(A)$.

So if I roll a die over and over, the proportion of them that give me a 6 over time should be $\frac{1}{6}$. So if we roll a dice 10000 times, we would expect one sixth of them to come up heads:

```
          n=10000

sixes=0
Sixvec<-vector(length = 0)

for (i in 1:n){
  roll=sample(1:6,1,replace=TRUE)
  if (roll==1){
    sixes=sixes+1
  }
Sixvec<-c(Sixvec, sixes/i)
}

plot(Sixvec, type="l", ylim=c(0,1))
abline(h=1/6, col="red", lty=2)
```

Definition A.2.4 A **random variable** is a function from sample space to an outcome set. For our purposes, this set of outcomes will always be $\mathbb{R}$. ◇

A **probability distribution** is, roughly speaking, a complete description of a random variable and the likelihood of each output. In the case of random variables with a finite number of possible outputs a **probability distribution table** is a convenient way of presenting this information.

Remark A.2.5 To check if something is a valid probability distribution, for any possible outcome x of X we must have:

- $0 \leq P(X = x) \leq 1$. This ensures all outcomes are valid probabilities.

- $\sum P(X = x) = 1$. The sum of the probabilities of all outcomes should be 100% of the outcomes

Example A.2.6 Poisoned apples. Snow White has a basket of 10 apples, 3 are poisoned. She is going to pick 4 apples at random to eat for some reason. Let X denote the number of poisoned apples she eats.

The probability distribution for X would be:

x	0	1	2	3
$P(X = x)$	$\dfrac{\binom{3}{0}\binom{7}{4}}{\binom{10}{4}}$	$\dfrac{\binom{3}{1}\binom{7}{3}}{\binom{10}{4}}$	$\dfrac{\binom{3}{2}\binom{7}{2}}{\binom{10}{4}}$	$\dfrac{\binom{3}{3}\binom{7}{1}}{\binom{10}{4}}$

equivalently:

x	0	1	2	3
$P(X = x)$	$\dfrac{35}{210}$	$\dfrac{105}{210}$	$\dfrac{63}{210}$	$\dfrac{7}{210}$

or:

x	0	1	2	3
$P(X = x)$	≈ 0.1667	0.5	0.3	≈ 0.0333

This can be seen by the following R simulation:

```
n=1000
poison = 3
notpoison = 7
eat = 4

numpoison = rep(0, n)
poisonvec = rep(c(1),each=poison)
notpoisonvec = rep(c(0),each=notpoison)
applevec = c(poisonvec, notpoisonvec)

for(i in 1:n){
  numpoison[i] = sum(sample(applevec, eat, replace = FALSE))
}
hist(numpoison)
```

$\square$

Definition A.2.7 Given a finite random variable X, its **expected value** is the predicted average outcome of experiments, and is computed:

$$E(X) = \sum P(X = x) \cdot x.$$

$\Diamond$

Note that the "Expected Value" may not be a value we actually expect, that is, may not be one of the outcomes, just an average outcome. We think of this as the outcomes of X, weighted by their likelihood, so the more likely outcomes contribute more than the less likely ones.

Example A.2.8 Recall Example A.2.6. The expected value of poisoned apples would be

$$E(X) = 0 \cdot \frac{35}{210} + 1 \cdot \frac{105}{210} + 2 \cdot \frac{63}{210} + 3 \cdot \frac{7}{210} = 1.2.$$

We can compute the mean of the previously simulated number of poisoned apples and visualize it:

```
hist(numpoison)
avgpoison = mean(numpoison)
abline(v = avgpoison, lwd = 2, lty = 2, col = "red")
print(avgpoison)
```

Be sure to run the simulation in Example A.2.6 first! $\square$

Appendix B

Simplex Pivoter

Instructions for use are as follows:

- First, enter in the number of variables and the number of bounds (the sizes of $\mathbf{c}, \mathbf{b}$ respectively.)

- In the cells generated below, fill in the entries.

- In the tableau generated below, click on an entry to pivot on that entry.

Index

Colophon

This book was authored in PreTeXt.